Springer Series in Statistics

Advisors:
P. Bickel, P. Diggle, S. Fienberg, K. Krickeberg,
I. Olkin, N. Wermuth, S. Zeger

Springer
New York
Berlin
Heidelberg
Barcelona
Hong Kong
London
Milan
Paris
Singapore
Tokyo

Springer Series in Statistics

Andersen/Borgan/Gill/Keiding: Statistical Models Based on Counting Processes.
Atkinson/Riani: Robust Diagnotstic Regression Analysis.
Berger: Statistical Decision Theory and Bayesian Analysis, 2nd edition.
Bolfarine/Zacks: Prediction Theory for Finite Populations.
Borg/Groenen: Modern Multidimensional Scaling: Theory and Applications
Brockwell/Davis: Time Series: Theory and Methods, 2nd edition.
Chen/Shao/Ibrahim: Monte Carlo Methods in Bayesian Computation.
David/Edwards: Annotated Readings in the History of Statistics.
Devroye/Lugosi: Combinatorial Methods in Density Estimation.
Efromovich: Nonparametric Curve Estimation: Methods, Theory, and Applications.
Fahrmeir/Tutz: Multivariate Statistical Modelling Based on Generalized Linear Models.
Farebrother: Fitting Linear Relationships: A History of the Calculus of Observations 1750-1900.
Federer: Statistical Design and Analysis for Intercropping Experiments, Volume I: Two Crops.
Federer: Statistical Design and Analysis for Intercropping Experiments, Volume II: Three or More Crops.
Fienberg/Hoaglin/Kruskal/Tanur (Eds.): A Statistical Model: Frederick Mosteller's Contributions to Statistics, Science and Public Policy.
Fisher/Sen: The Collected Works of Wassily Hoeffding.
Good: Permutation Tests: A Practical Guide to Resampling Methods for Testing Hypotheses, 2nd edition.
Gouriéroux: ARCH Models and Financial Applications.
Grandell: Aspects of Risk Theory.
Haberman: Advanced Statistics, Volume I: Description of Populations.
Hall: The Bootstrap and Edgeworth Expansion.
Härdle: Smoothing Techniques: With Implementation in S.
Hart: Nonparametric Smoothing and Lack-of-Fit Tests.
Hartigan: Bayes Theory.
Hedayat/Sloane/Stufken: Orthogonal Arrays: Theory and Applications.
Heyde: Quasi-Likelihood and its Application: A General Approach to Optimal Parameter Estimation.
Huet/Bouvier/Gruet/Jolivet: Statistical Tools for Nonlinear Regression: A Practical Guide with S-PLUS Examples.
Kolen/Brennan: Test Equating: Methods and Practices.
Kotz/Johnson (Eds.): Breakthroughs in Statistics Volume I.
Kotz/Johnson (Eds.): Breakthroughs in Statistics Volume II.
Kotz/Johnson (Eds.): Breakthroughs in Statistics Volume III.
Küchler/Sørensen: Exponential Families of Stochastic Processes.
Le Cam: Asymptotic Methods in Statistical Decision Theory.
Le Cam/Yang: Asymptotics in Statistics: Some Basic Concepts, 2nd edition.
Longford: Models for Uncertainty in Educational Testing.
Miller, Jr.: Simultaneous Statistical Inference, 2nd edition.

(continued after index)

Luc Devroye
Gábor Lugosi

Combinatorial Methods in Density Estimation

Springer

Luc Devroye
Computer Science Department
McGill University
Montreal, Quebec H3A 2K6
Canada

Gábor Lugosi
Universitat Pompeu Fabra
Facultat de Ciencies Economiques
Ramon Trias Fargas, 25-27
08005 Barcelona, Spain

Library of Congress Cataloging-in-Publication Data
Devroye, Luc.
Combinatorial methods in density estimation / Luc Devroye, Gábor Lugosi.
p. cm. — (Springer series in statistics)
Includes bibliographical references and index.
ISBN 0-387-95117-2 (alk. paper)
1. Distribution (Probability theory) 2. Estimation theory. 3. Combinatorial analysis.
I. Lugosi, Gábor. II. Title. III. Series.
QA273.6.D46 2000
512.2′4—dc21 00-058306

Printed on acid-free paper.

Production managed by Terry Kornak; manufacturing supervised by Jeffrey Taub.
Photocomposed copy prepared by the author.
Printed and bound by Edwards Brothers, Inc., Ann Arbor, MI.
Printed in the United States of America.

9 8 7 6 5 4 3 2 1

ISBN 0-387-95117-2 SPIN 10777879

Springer-Verlag New York Berlin Heidelberg
A member of BertelsmannSpringer Science+Business Media GmbH

to Bea and Arrate

to Dani, Birgit and Natasha

to Sid

Preface

In 1996 and 1997, we proposed a new method of bandwidth selection in density estimation that had universal optimality properties. This method had its foundations in the uniform convergence of empirical measures and in particular, in the Vapnik–Chervonenkis theory. Our approach was heavily influenced by a couple of papers in the late 1980s by Yannis Yatracos. We discovered later that the methodology extends to most popular density estimates and we decided to put down our ideas in this book. The goals of this text are to explain our methods, to illustrate some new fields of application in density estimation, and to present some new research results. This book has full sets of proofs and plenty of exercises, as well as a few research-level passages near the end. The title may shock the classical density estimation crowd that is perhaps not familiar with combinatorial counting methods. However, most arguments are quite elementary, and easy to absorb. The text is concise and moves at a brisk pace from idea to idea. Basic versions of all chapters were written in 1999 and sent back and forth by email between Barcelona and Montreal. The final versions were written in December 1999 in Montreal.

We thank all those people who influenced our work. They are all in the lists of references. We are especially grateful to our closest collaborators, Laci Györfi (Budapest) and Frederic Udina (Barcelona). Amalia Duch (Aguascalientes), Adam Krzyżak (Montreal), Balázs Kégl (Montreal), and Tamás Linder (Kingston) provided the necessary entertainment during our heavy writing schedule.

While we were writing our book, one of our best friends passed away. On September 1, 1999, Sid Yakowitz left us unexpectedly. He had a happy outlook on life, and always reminded us how relative all the things were that we were doing. He was very proud of his daughters, and loved his wife, Diana. It seems like yesterday when Sid told us how the coyotes around his house in Tucson kept preying on his pets. And the continuous stream of email jokes made it seem like he was in the office next door. We saw him for the last time in the Fall of 1997 at a nonparametrics workshop in Montreal, where he was seated in the front row of the lecture hall, and had a question for every speaker during the entire week, showing all the younger researchers how to stay alert and how to always question themselves and their work. To Sid, the storyteller, the father, the husband, our good friend, we dedicate this book.

Montreal and Barcelona
May 2000

Luc Devroye
Gábor Lugosi

Contents

Chapter 1
Introduction

A random variable X on $\mathbf{R}^d$ has a density f if, for all Borel sets A of $\mathbf{R}^d$, $\int_A f(x)\,dx = \mathbf{P}\{X \in A\}$. It thus serves as a tool for computing probabilities of sets. As it is a function that reveals the local concentration of probability mass, it may be used to visualize distributions of random variables. The statistician's problem, then, is to estimate f from an i.i.d. sample $X_1, \ldots, X_n$ drawn from f. A density estimate is simply a mapping $f_n : \mathbf{R}^d \times (\mathbf{R}^d)^n \to \mathbf{R}^d$ (we write $f_n(x; X_1, \ldots, X_n)$ or $f_n(x)$). It is the global closeness of f_n to f that interests us. The choice of a density estimate is governed by a number of factors, like consistency, smoothness, ease of computation, interpretability, flexibility, robustness, versatility, and optimality for certain criteria. The early work in the field approached the problem largely as a functional estimation problem: f was treated as any function, and tools from function approximation theory came to the rescue in the analysis of the performance of density estimates—Taylor series expansions played a key role, for example. The view we take in nonparametric density estimation is that f is largely unknown and that no assumptions can be made about its properties. Methods or properties that are valid for all densities are said to be universal. It is quite surprising that there are density estimates that can approximate any density f asymptotically in an appropriate sense. Simple histogram and kernel estimates have this property for example. But when people started studying rates of convergence, or error bounds for particular n, they invariably had to place conditions on the smoothness or boundedness of f. This was frustrating, because it was nearly impossible to test whether these assumptions were satisfied. Just try cooking up a test that decides whether the essential supremum of f is bounded, for example!

In 1996 and 1997, taking the kernel estimate as a prototype example, we showed in two papers how one can pick a kernel estimate very close to the optimal one, and have universal performance guarantees. The methods used in those papers were combinatorial in nature, and borrowed ideas from the work of Yatracos (1985, 1988) on minimum distance methods, and from the work of Vapnik and Chervonenkis (1971) on the uniform convergence of empirical probability measures over classes of sets. Our notes here explain these ideas in more detail, with all proofs given in full. Of course, the kernel estimate was only a prototype, so the view taken here is broader and essentially applicable to any density estimate and any dimension d. In the process, we will give new short proofs of the Vapnik–Chervonenkis inequality, and offer some fun calculations of shatter coefficients, an essential element of these inequalities. In fact, we are making it possible now to use the vast theory of Vapnik and Chervonenkis in density estimation, and hope that others will continue this work.

The 1990s has seen a rise in the use of concentration inequalities based on classical probability thory (Hoeffding's inequality, 1963), martingale methods (Azuma, 1967; McDiarmid, 1989), and inductive arguments (Talagrand, 1995, 1996; Ledoux, 1996). They are tremendously useful in density estimation and some of these deserve a special early treatment in these notes. The first part of this book (Chapters 2–4) contains introductory material necessary to make the book self-contained. It does not discuss density estimation, but gives an elementary introduction to some classical results on the concentration of measure, empirical processes, and offers the combinatorial foundation on which the rest of the book is built. The reader already familiar with these concepts may quickly scan these chapters.

The text is partitioned into virtually independent bite-sized pieces, each roughly digestable in one graduate student seminar. Chapters 5 through 8 provide a gentle introduction to the problem of density estimation and of the selection from a best estimate from a class of estimates. We hope that the examples provided there show how to select density estimates based on our L_1 minimum distance method. The analysis is reduced there to counting different intersections of finite sets of points with typically infinite classes of sets from $\mathbf{R}^d$.

Classical nonparametric density estimates such as the kernel estimate are not introduced until Chapter 9. In Chapters 10 and 11 we introduce the combinatorial tools for analyzing kernel estimates in which the bandwidth and other parameters are selected in a nearly optimal manner. The kernel estimate is dissected further in Chapters 12 (multivariate and multiparameter kernel estimates), 14 (transformed kernel estimates), 16 (automated choice of the order of a kernel), and 17 (superkernels). In Chapter 13, we take a look at the optimal choice of wavelet estimates, and in Chapter 15 we round out the book by introducing the minimax theory from the ground up. That chapter was influenced by the methods of Assouad, Birgé, and indirectly, LeCam.

A fundamental part of the material is placed in the exercise sections. Hard exercises are flagged. The references are kept to a minimum. Within a chapter, chapter numbers are dropped, so that within a chapter we will refer to Lemma 2, Theorem 3, and Section 4, for example.

Acknowledgment. These notes were written at McGill University and Universitat Pompeu Fabra. We thank both universities for providing us with wonderful facilities. Luc's research was supported by NSERC grant A3456. Gábor's research was supported by DGES grant PB96-0300.

§1.1. References

K. Azuma, "Weighted sums of certain dependent random variables," *Tohoku Mathematical Journal*, vol. 37, pp. 357–367, 1967.

L. Devroye and G. Lugosi, "A universally acceptable smoothing factor for kernel density estimation," *Annals of Statistics*, vol. 24, pp. 2499–2512, 1996.

L. Devroye and G. Lugosi, "Nonasymptotic universal smoothing factors, kernel complexity and Yatracos classes," *Annals of Statistics*, vol. 25, pp. 2626–2637, 1997.

W. Hoeffding, "Probability inequalities for sums of bounded random variables," *Journal of the American Statistical Association*, vol. 58, pp. 13–30, 1963.

M. Ledoux, "On Talagrand's deviation inequalities for product measures," *ESAIM: Probability and Statistics*, vol. 1, pp. 63–87, 1996.

C. McDiarmid, "On the method of bounded differences," in: *Surveys in Combinatorics 1989*, vol. 141, pp. 148–188, London Mathematical Society Lecture Notes Series, Cambridge University Press, Cambridge, 1989.

M. Talagrand, "Concentration of measure and isoperimetric inequalities in product spaces," *Institut des Hautes Etudes Scientifiques. Publications Mathématiques*, vol. 81, pp. 73–205, 1995.

M. Talagrand, "A new look at independence," *Annals of Probability*, vol. 24, pp. 1–34, 1996.

V. N. Vapnik and A. Ya. Chervonenkis, "On the uniform convergence of relative frequencies of events to their probabilities," *Theory of Probability and its Applications*, vol. 16, pp. 264–280, 1971.

Y. G. Yatracos, "Rates of convergence of minimum distance estimators and Kolmogorov's entropy," *Annals of Statistics*, vol. 13, pp. 768–774, 1985.

Y. G. Yatracos, "A note on L_1 consistent estimation," *Canadian Journal of Statistics*, vol. 16, pp. 283–292, 1988.

Chapter 2
Concentration Inequalities

Most of the methodology discussed in this book builds on elements of empirical process theory, certain concentration-of-measure inequalities, and simple combinatorial calculations. The aim of this and the following two chapters is to equip the reader with these simple tools. We keep the material at an elementary level, with additional information added in the exercises.

§2.1. Hoeffding's Inequality

Often we need inequalities to determine bounds for the difference between a random variable and its expected value. This chapter is a basic toolbox, where we gather some simple and useful results. We begin with *Markov's inequality*: for any nonnegative random variable X, and $t > 0$:

$$\mathbf{P}\{X \geq t\} \leq \frac{\mathbf{E}X}{t}.$$

From this, we deduce *Chebyshev's inequality*: if X is an arbitrary random variable and $t > 0$, then

$$\mathbf{P}\{|X - \mathbf{E}X| \geq t\} = \mathbf{P}\{|X - \mathbf{E}X|^2 \geq t^2\} \leq \frac{\mathbf{E}\{|X - \mathbf{E}X|^2\}}{t^2} = \frac{\mathbf{V}\{X\}}{t^2}.$$

As an example, we derive inequalities for $\mathbf{P}\{S_n - \mathbf{E}S_n \geq t\}$ with $S_n = \sum_{i=1}^n X_i$, where $X_1, \ldots, X_n$ are independent real-valued random variables. Chebyshev's inequality and independence immediately gives us

$$\mathbf{P}\{|S_n - \mathbf{E}S_n| \geq t\} \leq \frac{\mathbf{V}\{S_n\}}{t^2} = \frac{\sum_{i=1}^n \mathbf{V}\{X_i\}}{t^2}.$$

The meaning of this is perhaps better seen if we assume that the X_i's are i.i.d. Bernoulli (p) random variables (i.e., $\mathbf{P}\{X_i = 1\} = 1 - \mathbf{P}\{X_i = 0\} = p$), and normalize:

$$\mathbf{P}\left\{\left|\frac{1}{n}\sum_{i=1}^n X_i - p\right| \geq \epsilon\right\} \leq \frac{p(1-p)}{n\epsilon^2}.$$

To illustrate the weakness of this bound, let $\Phi(y) = \int_{-\infty}^{y} e^{-t^2/2}/\sqrt{2\pi}\,dt$ be the normal distribution function. The central limit theorem states that

$$\mathbf{P}\left\{\sqrt{\frac{n}{p(1-p)}}\left(\frac{1}{n}\sum_{i=1}^{n}X_i - p\right) \geq y\right\} \to 1-\Phi(y) \leq \frac{1}{\sqrt{2\pi}}\frac{e^{-y^2/2}}{y},$$

from which we would expect something like

$$\mathbf{P}\left\{\frac{1}{n}\sum_{i=1}^{n}X_i - p \geq \epsilon\right\} \approx e^{-n\epsilon^2/(2p(1-p))}.$$

Clearly, Chebyshev's inequality is off the mark. An improvement may be obtained by *Chernoff's bounding method.* By Markov's inequality, if s is an arbitrary positive number, then for any random variable X, and any $t > 0$:

$$\mathbf{P}\{X \geq t\} = \mathbf{P}\{e^{sX} \geq e^{st}\} \leq \frac{\mathbf{E}\{e^{sX}\}}{e^{st}}.$$

In Chernoff's method, we find an $s > 0$ that minimizes the upper bound or makes the upper bound small. In the case of a sum of independent random variables

$$\begin{aligned}\mathbf{P}\{S_n - \mathbf{E}S_n \geq t\} &\leq e^{-st}\mathbf{E}\left\{\exp\left(s\sum_{i=1}^{n}(X_i - \mathbf{E}X_i)\right)\right\} \\ &= e^{-st}\prod_{i=1}^{n}\mathbf{E}\{e^{s(X_i-\mathbf{E}X_i)}\} \quad \text{(by independence)}.\end{aligned}$$

Now the problem of finding tight bounds comes down to finding a good upper bound for the moment generating function of the random variables $X_i - \mathbf{E}X_i$. There are many ways of doing this. For bounded random variables perhaps the most elegant version is due to Hoeffding:

LEMMA 2.1. *Let X be a random variable with $\mathbf{E}X = 0$, $a \leq X \leq b$. Then for $s > 0$:*

$$\mathbf{E}\{e^{sX}\} \leq e^{s^2(b-a)^2/8}.$$

PROOF. Note that by convexity of the exponential function

$$e^{sx} \leq \frac{x-a}{b-a}e^{sb} + \frac{b-x}{b-a}e^{sa} \qquad \text{for } a \leq x \leq b.$$

Exploiting $\mathbf{E}X = 0$, and introducing the notation $p = -a/(b-a)$, we get

$$\begin{aligned}\mathbf{E}\{e^{sX}\} &\leq \frac{b}{b-a}e^{sa} - \frac{a}{b-a}e^{sb} \\ &= \left(1-p+pe^{s(b-a)}\right)e^{-ps(b-a)} \\ &\stackrel{\text{def}}{=} e^{\phi(u)},\end{aligned}$$

where $u = s(b-a)$ and $\phi(u) = -pu + \log(1 - p + pe^u)$. But by straightforward calculation it is easy to see that the derivative of ϕ is

$$\phi'(u) = -p + \frac{p}{p + (1-p)e^{-u}},$$

and therefore $\phi(0) = \phi'(0) = 0$. Moreover,

$$\phi''(u) = \frac{p(1-p)e^{-u}}{(p + (1-p)e^{-u})^2} \le 1/4.$$

Thus, by Taylor's theorem, for some $\theta \in [0, u]$:

$$\phi(u) = \phi(0) + u\phi'(0) + \frac{u^2}{2}\phi''(\theta) \le \frac{u^2}{8} = \frac{s^2(b-a)^2}{8}. \quad \square$$

Now we may directly plug this lemma into the bound obtained by Chernoff's method:

$$\begin{aligned}
\mathbf{P}\{S_n - \mathbf{E}S_n \ge t\} \\
&\le e^{-st} \prod_{i=1}^{n} e^{s^2(b_i - a_i)^2/8} \quad \text{(by Lemma 1)} \\
&= e^{-st} e^{s^2 \sum_{i=1}^{n} (b_i - a_i)^2/8} \\
&= e^{-2t^2/\sum_{i=1}^{n}(b_i - a_i)^2} \quad \text{(by choosing } s = 4t/\sum_{i=1}^{n}(b_i - a_i)^2 \text{).}
\end{aligned}$$

The result we have just derived is generally known as *Hoeffding's inequality*. For binomial random variables it was proved by Chernoff (1952) and Okamoto (1958). Summarizing, we have:

THEOREM 2.1 (HOEFFDING'S INEQUALITY). *Let $X_1, \ldots, X_n$ be independent bounded random variables such that X_i falls in the interval $[a_i, b_i]$ with probability one. Then for any $t > 0$, we have*

$$\mathbf{P}\{S_n - \mathbf{E}S_n \ge t\} \le e^{-2t^2/\sum_{i=1}^{n}(b_i - a_i)^2}$$

and

$$\mathbf{P}\{S_n - \mathbf{E}S_n \le -t\} \le e^{-2t^2/\sum_{i=1}^{n}(b_i - a_i)^2}.$$

If we specialize this to the binomial distribution, that is, when the X_i's are i.i.d. Bernoulli (p), we get

$$\mathbf{P}\{S_n/n - p \ge \epsilon\} \le e^{-2n\epsilon^2},$$

which is just the kind of inequality we hoped for.

§2.2. An Inequality for the Expected Maximal Deviation

Sometimes we need upper bounds for the expected value of a maximum of centralized binomial random variables. Such bounds may be derived by observing that for any nonnegative random variable X:

$$\mathbf{E}X = \int_0^\infty \mathbf{P}\{X \geq t\}\, dt,$$

and using Theorem 1. Sharper bounds result by combining Lemma 1 with the following simple result:

LEMMA 2.2. *Let $\sigma > 0$, $n \geq 2$, and let $Y_1, \ldots, Y_n$ be real-valued random variables such that for all $s > 0$ and $1 \leq i \leq n$, $\mathbf{E}\{e^{sY_i}\} \leq e^{s^2\sigma^2/2}$. Then*

$$\mathbf{E}\left\{\max_{i \leq n} Y_i\right\} \leq \sigma\sqrt{2\ln n}.$$

If, in addition, $\mathbf{E}\{e^{s(-Y_i)}\} \leq e^{s^2\sigma^2/2}$ for every $s > 0$ and $1 \leq i \leq n$, then for any $n \geq 1$,

$$\mathbf{E}\left\{\max_{i \leq n} |Y_i|\right\} \leq \sigma\sqrt{2\ln(2n)}.$$

PROOF. By Jensen's inequality, for all $s > 0$:

$$\begin{aligned} e^{s\mathbf{E}\{\max_{i \leq n} Y_i\}} &\leq \mathbf{E}\{e^{s \max_{i \leq n} Y_i}\} \\ &= \mathbf{E}\left\{\max_{i \leq n} e^{sY_i}\right\} \\ &\leq \sum_{i=1}^n \mathbf{E}\{e^{sY_i}\} \\ &\leq n e^{s^2\sigma^2/2}. \end{aligned}$$

Thus,

$$\mathbf{E}\left\{\max_{i \leq n} Y_i\right\} \leq \frac{\ln n}{s} + \frac{s\sigma^2}{2},$$

and taking $s = \sqrt{2\ln n/\sigma^2}$ yields the first inequality. Finally, note that $\max_{i \leq n} |Y_i| = \max(Y_1, -Y_1, \ldots, Y_n, -Y_n)$ and apply the first inequality to prove the second. □

§2.3. The Bounded Difference Inequality

In this section we give a powerful extension of Hoeffding's inequality to general functions of independent random variables.

Let A be some set, and assume that a function $g : A^n \to \mathbf{R}$ satisfies the *bounded difference assumption*

$$\sup_{\substack{x_1,\ldots,x_n,\\ x_i' \in A}} |g(x_1,\ldots,x_n) - g(x_1,\ldots,x_{i-1},x_i',x_{i+1},\ldots,x_n)| \le c_i, \quad 1 \le i \le n.$$

In other words, we assume that if we change the ith variable of g while keeping all the others fixed, then the value of the function does not change by more than c_i.

We derive inequalities for the difference between $g(X_1,\ldots,X_n)$ and its expected value when $X_1,\ldots,X_n$ are arbitrary independent random variables taking values in A. Sometimes we will write g instead of $g(X_1,\ldots,X_n)$ whenever it does not cause any confusion.

THEOREM 2.2 (THE BOUNDED DIFFERENCE INEQUALITY). *Under the bounded difference assumption above, for all $t > 0$,*

$$\mathbf{P}\{g(X_1,\ldots,X_n) - \mathbf{E}g(X_1,\ldots,X_n) \ge t\} \le e^{-2t^2/\sum_{i=1}^n c_i^2},$$

and

$$\mathbf{P}\{\mathbf{E}g(X_1,\ldots,X_n) - g(X_1,\ldots,X_n) \ge t\} \le e^{-2t^2/\sum_{i=1}^n c_i^2}.$$

McDiarmid (1989) proved this inequality using martingale techniques, which we reproduce here. We recall the elementary fact that if X and Y are arbitrary bounded random variables, then $\mathbf{E}\{XY\} = \mathbf{E}\{\mathbf{E}\{XY|Y\}\} = \mathbf{E}\{Y\mathbf{E}\{X|Y\}\}$. The proof of Theorem 2 uses the following straightforward extension of Lemma 1:

LEMMA 2.3. *Let V and Z be random variables such that $\mathbf{E}\{V|Z\} = 0$ with probability one, and for some function h and constant $c \ge 0$:*

$$h(Z) \le V \le h(Z) + c.$$

Then for all $s > 0$:

$$\mathbf{E}\{e^{sV}|Z\} \le e^{s^2c^2/8}.$$

PROOF OF THEOREM 2. Introduce the notation $V = g - \mathbf{E}g$, and define

$$V_i = \mathbf{E}\{g|X_1,\ldots,X_i\} - \mathbf{E}\{g|X_1,\ldots,X_{i-1}\}, \qquad i = 1,\ldots,n.$$

Clearly, $V = \sum_{i=1}^n V_i$. Also introduce the random variables

$$H_i(X_1,\ldots,X_i) = \mathbf{E}\{g(X_1,\ldots,X_n)|X_1,\ldots,X_i\}.$$

Then, denoting the distribution of X_i by F_i for $i = 1,\ldots,n$:

$$V_i = H_i(X_1,\ldots,X_i) - \int H_i(X_1,\ldots,X_{i-1},x)F_i(dx).$$

Define the random variables

$$W_i = \sup_u \left(H_i(X_1, \ldots, X_{i-1}, u) - \int H_i(X_1, \ldots, X_{i-1}, x) F_i(dx) \right),$$

and

$$Z_i = \inf_v \left(H_i(X_1, \ldots, X_{i-1}, v) - \int H_i(X_1, \ldots, X_{i-1}, x) F_i(dx) \right).$$

Clearly, $Z_i \le V_i \le W_i$ with probability one, and also

$$W_i - Z_i = \sup_u \sup_v \left(H_i(X_1, \ldots, X_{i-1}, u) - H_i(X_1, \ldots, X_{i-1}, v) \right) \le c_i,$$

by the bounded difference assumption. Therefore, we may apply Lemma 3 to obtain, for all $i = 1, \ldots, n$:

$$\mathbf{E}\{e^{sV_i} | X_1, \ldots, X_{i-1}\} \le e^{s^2 c_i^2 / 8}.$$

Finally, by Chernoff's bound, for any $s > 0$:

$$\begin{aligned}
&\mathbf{P}\{g - \mathbf{E}g \ge t\} \\
&\quad \le \frac{\mathbf{E}\left\{e^{s \sum_{i=1}^n V_i}\right\}}{e^{st}} = \frac{\mathbf{E}\left\{e^{s \sum_{i=1}^{n-1} V_i} \mathbf{E}\{e^{sV_n} | X_1, \ldots, X_{n-1}\}\right\}}{e^{st}} \\
&\quad \le e^{s^2 c_n^2 / 8} \frac{\mathbf{E}\left\{e^{s \sum_{i=1}^{n-1} V_i}\right\}}{e^{st}} \\
&\quad \le e^{-st} e^{s^2 \sum_{i=1}^n c_i^2 / 8}
\end{aligned}$$

by repeating the same argument n times. Choosing $s = 4t \Big/ \sum_{i=1}^n c_i^2$ proves the first inequality. The proof of the second inequality is similar. □

§2.4. Examples

First of all, note that if

$$g(X_1, \ldots, X_n) = \sum_{i=1}^n X_i,$$

where $X_i \in [a_i, b_i]$, then the bounded difference assumption is satisfied with $c_i = b_i - a_i$, and therefore Hoeffding's inequality is a special case of Theorem 2.

Next assume that $X_1, \ldots, X_n$ are i.i.d. random variables taking their values in some set $\mathcal{X}$, and let $\mathcal{A}$ be a collection of subsets of $\mathcal{X}$. Let μ denote the distribution of X_1, that is, $\mu(A) = \mathbf{P}\{X_1 \in A\}$, $A \in \mathcal{A}$, and let μ_n denote the empirical distribution:

$$\mu_n(A) = \frac{1}{n} \sum_{i=1}^n I_{\{X_n \in A\}}.$$

The quantity of interest is the largest deviation of μ_n from μ over $\mathcal{A}$:

$$g(X_1, \ldots, X_n) = \sup_{A \in \mathcal{A}} |\mu_n(A) - \mu(A)|.$$

In the next chapter we provide sufficient conditions on $\mathcal{A}$ under which this supremum converges to zero in probability. But regardless of what $\mathcal{A}$ is, by changing one X_i, g can change by at most $1/n$. Therefore,

$$\mathbf{P}\{|g - \mathbf{E}g| \geq \epsilon\} \leq 2e^{-2n\epsilon^2}.$$

One implication is that $g \to 0$ in probability if and only if $g \to 0$ almost surely.

Finally, let $X_1, \ldots, X_n$ be i.i.d. samples drawn according to some density f on the real line. The density is estimated by the kernel estimate

$$f_n(x) = \frac{1}{nh} \sum_{i=1}^{n} K\left(\frac{x - X_i}{h}\right),$$

where $h > 0$ is a smoothing parameter, and K is a nonnegative function with $\int K = 1$. The performance of the estimate is measured by the L_1 error

$$g(X_1, \ldots, X_n) = \int |f_n(x) - f(x)|\, dx.$$

It is easy to see that

$$\begin{aligned} &|g(x_1, \ldots, x_n) - g(x_1, \ldots, x_i', \ldots, x_n)| \\ &\qquad \leq \frac{1}{nh} \int \left| K\left(\frac{x - x_i}{h}\right) - K\left(\frac{x - x_i'}{h}\right)\right| dx \leq \frac{2}{n}, \end{aligned}$$

so without further work we get

$$\mathbf{P}\{|g - \mathbf{E}g| \geq \epsilon\} \leq 2e^{-n\epsilon^2/2}.$$

§2.5. Bibliographic Remarks

The theory of exponential tail inequalities for sums of independent random variables was pioneered by Bernstein (1946), Craig (1933), Chernoff (1952), Okamoto (1958), Bennett (1962), Hoeffding (1963), and others. Hoeffding (1963) and Azuma (1967) observed that Chernoff's method may be extended to martingales. However, the significance of this simple extension was not recognized until the late 1980s, when it triggered a revolution in many applications. Milman and Schechtman (1986), Rhee and Talagrand (1987), and Shamir and Spencer (1987) were among the first to use the martingale method to prove concentration results. The bounded difference inequality given here was formulated by McDiarmid (1989). There are several other methods to prove concentration. The first concentration result seems to have been published in the area of information theory by Ahlswede, Gács, and Körner (1976). It was perfected and greatly simplified by Marton (1986, 1996, 1997). See also Dembo (1997) and Massart (1998). A third method, which we might call the "induction method," was developed in the work of Talagrand (1988, 1989, 1990, 1991a,b, 1993a,b, 1994, 1995, 1996a,b). A fourth proof method, based on logarithmic Sobolev inequalities, was introduced by Ledoux (1996a,b) (see also Massart (1999), and Boucheron, Lugosi, and Massart (2000)). In density estimation, concentration inequalities were introduced by Devroye (1991). For recent surveys on concentration inequalities we refer to McDiarmid (1998) and Talagrand (1996a).

§2.6. Exercises

Exercise 2.1 (CHEBYSHEV-CANTELLI INEQUALITY). Prove the following one-sided improved version of Chebyshev's inequality: For any random variable X with $\mathbf{E}X^2 < \infty$ and $t > 0$:

$$\mathbf{P}\{X - \mathbf{E}X \ge t\} \le \frac{\mathbf{V}\{X\}}{\mathbf{V}\{X\} + t^2}.$$

HINT: Write $t = \mathbf{E}\{t - X\} \le \mathbf{E}\{(t - X)\mathbf{1}_{[X<t]}\}$, and apply the Cauchy–Schwarz inequality.

Exercise 2.2 (BENNETT'S INEQUALITY). Let $X_1, \ldots, X_n$ be independent real-valued random variables with zero mean, and assume that $|X_i| \le c$ with probability one. Let $\sigma^2 = (1/n)\sum_{i=1}^n \mathbf{V}\{X_i\}$. Prove that for any $t > 0$:

$$\mathbf{P}\{S_n > t\} \le \exp\left(-\frac{n\sigma^2}{c^2} h\left(\frac{ct}{n\sigma^2}\right)\right),$$

where the function h is defined by $h(u) = (1 + u)\log(1 + u) - u$ for $u \ge 0$. HINT: Use Chernoff's method as in the proof of Hoeffding's inequality. Write

$$\mathbf{E}\{e^{sX_i}\} = 1 + s\mathbf{E}\{X_i\} + \sum_{r=2}^{\infty} \frac{s^r \mathbf{E}\{X_i^r\}}{r!} = 1 + s^2 \mathbf{V}\{X_i\} F_i \le e^{s^2 \mathbf{V}\{X_i\} F_i}$$

with $F_i = \sum_{r=2}^{\infty} s^{r-2}\mathbf{E}\{X_i^r\}/(r!\mathbf{V}\{X_i\})$. Use the boundedness of the X_i's to show that $\mathbf{E}\{X_i^r\} \le c^{r-2}\mathbf{V}\{X_i\}$, which implies $F_i < (e^{sc} - 1 - sc)/(sc)^2$. Choose the s which minimizes the obtained upper bound for the tail probability.

Exercise 2.3 (BERNSTEIN'S INEQUALITY). Show that under the conditions of the previous exercise, for any $t > 0$,

$$\mathbf{P}\{S_n > t\} \le \exp\left(-\frac{t^2}{2n\sigma^2 + 2ct/3}\right).$$

HINT: The result follows from Bennett's inequality and the inequality $h(u) \geq u^2/(2+2u/3)$, $u \geq 0$.

Exercise 2.4 Let $X_1, \ldots, X_n$ be independent random variables, taking their values from $[0,1]$. Show that if $m = \mathbf{E}S_n$, then for any $m \leq t \leq n$:

$$\mathbf{P}\{S_n \geq t\} \leq \left(\frac{m}{t}\right)^t \left(\frac{n-m}{n-t}\right)^{n-t}.$$

Exercise 2.5 (CONTINUATION). Use the result of the previous exercise to conclude that:

A.

$$\mathbf{P}\{S_n \geq t\} \leq \left(\frac{m}{t}\right)^t e^{t-m}.$$

B. For all $\epsilon > 0$:

$$\mathbf{P}\{S_n \geq m(1+\epsilon)\} \leq e^{-mh(\epsilon)},$$

where h is the function defined in Exercise 2. Use the bound for h given in Exercise 3 to obtain a simpler bound.

C.

$$\mathbf{P}\{S_n \leq m(1-\epsilon)\} \leq e^{-m\epsilon^2/2}.$$

See Boucheron, Lugosi, and Massart (2000) for a much more general version of these inequalities.

Exercise 2.6 Assume that the function $g : A^n \to \mathbf{R}$ satisfies the bounded difference assumption, and let $X_1, \ldots, X_n$ be independent random variables taking values in A. Show that

$$\mathbf{V}\{g(X_1, \ldots, X_n)\} \leq 1/4 \sum_{i=1}^{n} c_i^2$$

(Devroye, 1991). HINT: Write $g - \mathbf{E}g$ as a sum of bounded martingale differences as in the proof of Theorem 2. Use the martingale property to show that $\mathbf{V}\{g\} = \sum_{i=1}^n \mathbf{E}V_i^2$, and bound each term of the sum using the boundedness shown in the proof of Theorem 3.

Exercise 2.7 Assume in addition to the conditions of the previous exercise that each X_i is a binary random variable, with $\mathbf{P}\{X_i = 1\} = 1 - \mathbf{P}\{X_i = 0\} = p_i$. Improve the bound of the previous exercise to

$$\mathbf{V}\{g(X_1, \ldots, X_n)\} \leq \sum_{i=1}^{n} p_i(1-p_i)c_i^2$$

(Kim, 1995). HINT: Use Jensen's inequality to prove that if h is a convex function, then for each $i = 1, \ldots, n$:

$$\mathbf{E}\{h(V_i)|X_1, \ldots, X_{i-1}\} \leq \mathbf{E}\{p_i h((1-p_i)\gamma_i) + (1-p_i)h(-p_i\gamma_i)|X_1, \ldots, X_{i-1}\},$$

where $\gamma_i = \beta_i - \alpha_i$ with

$$\alpha_i = g(X_1, \ldots, X_{i-1}, 0, X_{i+1}, \ldots, X_n)$$

and

$$\beta_i = g(X_1, \ldots, X_{i-1}, 1, X_{i+1}, \ldots, X_n),$$

and use it with $h(u) = u^2$. Can you improve Theorem 2 in this special case by using the above inequality with $h(u) = e^{su}$?

Exercise 2.8 Consider a vector of random variables $X = (X_1, \ldots, X_n)$ taking its value in the n-fold product A^n of some set A. Assume that $X_1, \ldots, X_n$ are independent. The Hamming distance $d(x, y)$ between the vectors $x, y \in A^n$ is defined as the number of components in which x and y differ. If $B \subset A^n$ is an arbitrary (measurable) set, then introduce

$$d(B, x) = \min_{y \in B} d(x, y).$$

Use the bounded difference inequality to derive that for all $\epsilon > 0$:

$$\mathbf{P}\{d(B, X) \geq n\epsilon\} \leq e^{-2n\left(\epsilon - \sqrt{(1/2n)\log(1/P(B))}\right)^2}$$

whenever

$$\epsilon \geq \sqrt{\frac{1}{2n} \log \frac{1}{P(B)}},$$

where $P(B) = \mathbf{P}\{X \in B\}$. See Talagrand (1996b) for more information.

Exercise 2.9 CONTINUATION. Define the Hamming distance between two sets $B, C \subset A^n$ by

$$d(B, C) = \min_{x \in B, y \in C} d(x, y).$$

Show that for all $B, C \subset A^n$:

$$d(B, C) \leq \sqrt{\frac{n}{2} \log \frac{1}{P(B)}} + \sqrt{\frac{n}{2} \log \frac{1}{P(C)}},$$

and, in fact, this property is equivalent with the statement of the previous exercise (Marton, 1996).

§2.7. References

R. Ahlswede, P. Gács, and J. Körner, "Bounds on conditional probabilities with applications in multi-user communication," *Zeitschrift für Wahrscheinlichkeitstheorie und verwandte Gebiete*, vol. 34, pp. 157–177, 1976 (correction in vol. 39, pp. 353–354, 1977).

K. Azuma, "Weighted sums of certain dependent random variables," *Tohoku Mathematical Journal*, vol. 68, pp. 357–367, 1967.

G. Bennett, "Probability inequalities for the sum of independent random variables," *Journal of the American Statistical Association*, vol. 57, pp. 33–45, 1962.

S. N. Bernstein, *The Theory of Probabilities*, Gastehizdat, Moscow, 1946.

S. Boucheron, G. Lugosi, and P. Massart, "A sharp concentration inequality with applications in random combinatorics and learning," *Random Structures and Algorithms*, 1999. To appear.

H. Chernoff, "A measure of asymptotic efficiency of tests of a hypothesis based on the sum of observations," *Annals of Mathematical Statistics*, vol. 23, pp. 493–507, 1952.

C. Craig, "On the Tchebychef inequality of Bernstein," *Annals of Mathematical Statistics*, vol. 4, pp. 94–102, 1933.

A. Dembo, "Information inequalities and concentration of measure," *Annals of Probability*, vol. 25, pp. 927–939, 1997.

L. Devroye, "Exponential inequalities in nonparametric estimation," in: *Nonparametric Functional Estimation and Related Topics* (edited by G. Roussas), pp. 31–44, Kluwer Academic, Dordrecht, 1991.

W. Hoeffding, "Probability inequalities for sums of bounded random variables," *Journal of the American Statistical Association*, vol. 58, pp. 13–30, 1963.

J. H. Kim, "The Ramsey number $R(3,t)$ has order of magnitude $t^2/\log t$," *Random Structures and Algorithms*, vol. 7, pp. 173–207, 1995.

M. Ledoux, "On Talagrand's deviation inequalities for product measures," *ESAIM: Probability and Statistics*, vol. 1, pp. 63–87, 1996a.

M. Ledoux, "Isoperimetry and gaussian analysis," in: *Lectures on Probability Theory and Statistics* (edited by P. Bernard), pp. 165–294, Ecole d'Eté de Probabilités de St-Flour XXIV-1994, 1996b.

K. Marton, "A simple proof of the blowing-up lemma," *IEEE Transactions on Information Theory*, vol. 32, pp. 445–446, 1986.

K. Marton, "Bounding $\bar{d}$-distance by informational divergence: a way to prove measure concentration," *Annals of Probability*, vol. 24, pp. 857–866, 1996.

K. Marton, "A measure concentration inequality for contracting Markov chains," *Geometric and Functional Analysis*, vol. 6, pp. 556–571, 1996. Erratum: vol. 7, pp. 609–613, 1997.

P. Massart, "Optimal constants for Hoeffding-type inequalities," Mathématiques Université de Paris-Sud, Report 98.86, 1998.

P. Massart, "About the constants in Talagrand's concentration inequalities for empirical processes," *Annals of Probability*, 1999. To appear.

C. McDiarmid, "On the method of bounded differences," in: *Surveys in Combinatorics 1989*, pp. 148–188, Cambridge University Press, Cambridge, 1989.

C. McDiarmid, "Concentration," in: *Probabilistic Methods for Algorithmic Discrete Mathematics* (edited by M. Habib, C. McDiarmid, J. Ramirez-Alfonsin and B. Reed), pp. 195–248, Springer-Verlag, New York, 1998.

V. Milman and G. Schechtman, *Asymptotic Theory of Finite-Dimensional Normed Spaces*, Springer-Verlag, New York, 1986.

M. Okamoto, "Some inequalities relating to the partial sum of binomial probabilities," *Annals of the Institute of Statistical Mathematics*, vol. 10, pp. 29–35, 1958.

W. Rhee and M. Talagrand, "Martingales, inequalities, and NP-complete problems," *Mathematics of Operations Research*, vol. 12, pp. 177–181, 1987.

E. Shamir and J. Spencer, "Sharp concentration of the chromatic number on random graphs $G_{n,p}$," *Combinatorica*, vol. 7, pp. 374–384, 1987.

M. Talagrand, "An isoperimetric theorem on the cube and the Kintchine–Kahane inequalities," *Proceedings of the American Mathematical Society*, vol. 104, pp. 905–909, 1988.

M. Talagrand, "Isoperimetry and integrability of the sum of independent Banach-space valued random variables," *Annals of Probability*, vol. 17, pp. 1546–1570, 1989.

M. Talagrand, "Sample unboundedness of stochastic processes under increment conditions," *Annals of Probability*, vol. 18, pp. 1–49, 1990.

M. Talagrand, "A new isoperimetric inequality and the concentration of measure phenomenon," in: *Geometric Aspects of Functional Analysis* (edited by J. Lindenstrauss and V. D. Milman), vol. 146, pp. 4–137, Lecture Notes in Mathematics, Springer-Verlag, Berlin, 1991a.

M. Talagrand, "A new isoperimetric inequality for product measure and the tails of sums of independent random variables," *Geometric and Functional Analysis*, vol. 1, pp. 211–223, 1991b.

M. Talagrand, "Isoperimetry, logarithmic Sobolev inequalities on the discrete cube, and Margulis' graph connectivity theorem," *Geometric and Functional Analysis*, vol. 3, pp. 295–314, 1993a.

M. Talagrand, "New Gaussian estimates for enlarged balls," *Geometric and Functional Analysis*, vol. 3, pp. 502–526, 1993b.

M. Talagrand, "Sharper bounds for Gaussian and empirical processes," *Annals of Probability*, vol. 22, pp. 28–76, 1994.

M. Talagrand, "Concentration of measure and isoperimetric inequalities in product spaces," *Institut des Hautes Études Scientifiques. Publications Mathématiques*, vol. 81, pp. 73–205, 1995.

M. Talagrand, "New concentration inequalities in product spaces," *Inventiones Mathematicae*, vol. 126, pp. 505–563, 1996a.

M. Talagrand, "A new look at independence," *Annals of Probability*, vol. 24, pp. 1–34, 1996b.

Chapter 3
Uniform Deviation Inequalities

§3.1. The Vapnik–Chervonenkis Inequality

This chapter is devoted to some basic inequalities that bound the maximal difference between probabilities and relative frequencies over a class of events. The bounds will be key tools in our study of density estimates. Let $X_1, \ldots, X_n$ be i.i.d. random variables taking values in $\mathbf{R}^d$ with common distribution

$$\mu(A) = \mathbf{P}\{X_1 \in A\} \ (A \subset \mathbf{R}^d).$$

Define the empirical distribution

$$\mu_n(A) = \frac{1}{n} \sum_{i=1}^{n} \mathbf{1}_{[X_i \in A]} \ (A \subset \mathbf{R}^d).$$

Consider a class $\mathcal{A}$ of subsets of $\mathbf{R}^d$. Our main concern here is the behavior of $\sup_{A \in \mathcal{A}} |\mu_n(A) - \mu(A)|$. We saw in the previous chapter that a simple consequence of the bounded difference inequality (Theorem 2.2) is that

$$\mathbf{P}\left\{ \left| \sup_{A \in \mathcal{A}} |\mu_n(A) - \mu(A)| - \mathbf{E}\left\{ \sup_{A \in \mathcal{A}} |\mu_n(A) - \mu(A)| \right\} \right| > t \right\} \le 2e^{-2nt^2}$$

for any n and $t > 0$. This shows that for any class $\mathcal{A}$, the maximal deviation is sharply concentrated around its mean. In the rest of this chapter we derive inequalities for the expected value, in terms of certain combinatorial quantities related to $\mathcal{A}$. The first such quantity is the *Vapnik–Chervonenkis shatter coefficient*, defined by

$$\mathbf{S}_{\mathcal{A}}(n) = \max_{x_1, \ldots, x_n \subset \mathbf{R}^d} |\{\{x_1, \ldots, x_n\} \cap A; A \in \mathcal{A}\}|.$$

In words, $\mathbf{S}_{\mathcal{A}}(n)$ is the maximal number of different subsets of a set of n points which can be obtained by intersecting it with elements of $\mathcal{A}$. We have the following version of a classical result of Vapnik and Chervonenkis:

THEOREM 3.1.

$$\mathbf{E}\left\{\sup_{A\in\mathcal{A}}|\mu_n(A)-\mu(A)|\right\}\le 2\sqrt{\frac{\log 2\mathbf{S}_{\mathcal{A}}(n)}{n}}.$$

PROOF. Introduce $X_1',\ldots,X_n'$, an independent copy of $X_1,\ldots,X_n$. Also, define n i.i.d. sign variables $\sigma_1,\ldots,\sigma_n$ such that $\mathbf{P}\{\sigma_1=-1\}=\mathbf{P}\{\sigma_1=1\}=1/2$, independent of $X_1,X_1',\ldots,X_n,X_n'$. Then, denoting $\mu_n'(A)=(1/n)\sum_{i=1}^n \mathbf{1}_{[X_i'\in A]}$, we may write

$$\begin{aligned}
&\mathbf{E}\left\{\sup_{A\in\mathcal{A}}|\mu_n(A)-\mu(A)|\right\}\\
&=\mathbf{E}\left\{\sup_{A\in\mathcal{A}}\left|\mathbf{E}\{\mu_n(A)-\mu_n'(A)|X_1,\ldots,X_n\}\right|\right\}\\
&\le\mathbf{E}\left\{\sup_{A\in\mathcal{A}}\mathbf{E}\{\left|\mu_n(A)-\mu_n'(A)\right|\,|X_1,\ldots,X_n\}\right\}\\
&\quad\text{(by Jensen's inequality)}\\
&\le\mathbf{E}\left\{\sup_{A\in\mathcal{A}}\left|\mu_n(A)-\mu_n'(A)\right|\right\}\\
&\quad\text{(since }\sup\mathbf{E}(\cdot)\le\mathbf{E}\sup(\cdot))\\
&=\frac{1}{n}\mathbf{E}\left\{\sup_{A\in\mathcal{A}}\left|\sum_{i=1}^n\sigma_i(\mathbf{1}_{[X_i\in A]}-\mathbf{1}_{[X_i'\in A]})\right|\right\}\\
&\quad\text{(because }X_1,X_1',\ldots,X_n,X_n'\text{ are i.i.d.)}\\
&=\frac{1}{n}\mathbf{E}\left\{\mathbf{E}\left\{\sup_{A\in\mathcal{A}}\left|\sum_{i=1}^n\sigma_i(\mathbf{1}_{[X_i\in A]}-\mathbf{1}_{[X_i'\in A]})\right|\,\middle|\,X_1,X_1',\ldots,X_n,X_n'\right\}\right\}.
\end{aligned}$$

Now because of the independence of the σ_i's of the rest of the variables, we may fix the values of $X_1=x_1,X_1'=x_1',\ldots,X_n=x_n,X_n'=x_n'$, and investigate

$$\mathbf{E}\left\{\sup_{A\in\mathcal{A}}\left|\sum_{i=1}^n\sigma_i(\mathbf{1}_{[x_i\in A]}-\mathbf{1}_{[x_i'\in A]})\right|\right\}.$$

Denote by $\widehat{\mathcal{A}}\subset\mathcal{A}$ a collection of sets such that any two sets in $\widehat{\mathcal{A}}$ have different intersections with the set $\{x_1,x_1',\ldots,x_n,x_n'\}$, and every possible intersection is represented once. Thus, $|\widehat{\mathcal{A}}|\le\mathbf{S}_{\mathcal{A}}(2n)$, and

$$\begin{aligned}
&\mathbf{E}\left\{\sup_{A\in\mathcal{A}}\left|\sum_{i=1}^n\sigma_i(\mathbf{1}_{[x_i\in A]}-\mathbf{1}_{[x_i'\in A]})\right|\right\}\\
&\qquad=\mathbf{E}\left\{\max_{A\in\widehat{\mathcal{A}}}\left|\sum_{i=1}^n\sigma_i(\mathbf{1}_{[x_i\in A]}-\mathbf{1}_{[x_i'\in A]})\right|\right\}.
\end{aligned}$$

Observing that each $\sigma_i(\mathbf{1}_{[x_i \in A]} - \mathbf{1}_{[x_i' \in A]})$ has zero mean and takes values in $[-1, 1]$, we obtain from Lemma 2.1 that for any $s > 0$:

$$\mathbf{E}\left\{e^{s\sum_{i=1}^n \sigma_i(\mathbf{1}_{[x_i\in A]}-\mathbf{1}_{[x_i'\in A]})}\right\}$$

$$= \prod_{i=1}^n \mathbf{E}\left\{e^{s\sigma_i(\mathbf{1}_{[x_i\in A]}-\mathbf{1}_{[x_i'\in A]})}\right\} \leq e^{ns^2/2}.$$

Since the distribution of $\sigma_i(\mathbf{1}_{[x_i \in A]} - \mathbf{1}_{[x_i' \in A]})$ is symmetric, Lemma 2.2 immediately implies that

$$\mathbf{E}\left\{\max_{A\in\widehat{\mathcal{A}}}\left|\sum_{i=1}^n \sigma_i(\mathbf{1}_{[x_i\in A]} - \mathbf{1}_{[x_i'\in A]})\right|\right\} \leq \sqrt{2n\log 2\mathbf{S}_{\mathcal{A}}(2n)}.$$

Conclude by observing that $\mathbf{S}_{\mathcal{A}}(2n) \leq \mathbf{S}_{\mathcal{A}}(n)^2$. □

§3.2. Covering Numbers and Chaining

Next we derive a refined version of the Vapnik–Chervonenkis theorem. The new bound involves some geometric and combinatorial quantities related to the class $\mathcal{A}$. Consider a pair of bit vectors $b = (b_1, \ldots, b_n)$ and $c = (c_1, \ldots, c_n)$ from $\{0,1\}^n$, and define their distance by

$$\rho(b,c) = \sqrt{\frac{1}{n}\sum_{i=1}^n \mathbf{1}_{[b_i \neq c_i]}}.$$

Thus, $\rho(b,c)$ is just the square root of the normalized Hamming distance between b and c. Observe that ρ may also be considered as the normalized Euclidean distance between the corners of the hypercube $[0,1]^n \subset \mathbf{R}^n$, and therefore it is indeed a distance.

Now let $B \subset \{0,1\}^n$ be any set of bit vectors, and define a *cover* of radius $r > 0$ as a set $B_r \subset \{0,1\}^n$ such that for any $b \in B$ there exists a $c \in B_r$ such that $\rho(b,c) \leq r$. The *covering number* $N(r, B)$ is the cardinality of the smallest cover of radius r.

A class $\mathcal{A}$ of subsets of $\mathbf{R}^d$ and a set of n points $x_1^n = \{x_1, \ldots, x_n\} \subset \mathbf{R}^d$ define a set of bit vectors by

$$\mathcal{A}(x_1^n) = \{b = (b_1, \ldots, b_n) \in \{0,1\}^n : \exists A \in \mathcal{A} : b_i = \mathbf{1}_{[x_i \in A]},\ i = 1, \ldots, n\}.$$

That is, every bit vector $b \in \mathcal{A}(x_1^n)$ describes the intersection of $\{x_1, \ldots, x_n\}$ with a set A in $\mathcal{A}$. We have the following:

THEOREM 3.2.

$$\mathbf{E}\left\{\sup_{A\in\mathcal{A}}|\mu_n(A) - \mu(A)|\right\} \leq \frac{24}{\sqrt{n}} \max_{x_1,\ldots,x_n\in\mathbf{R}^d} \int_0^1 \sqrt{\log 2N(r, \mathcal{A}(x_1^n))}\, dr.$$

The theorem implies that $\mathbf{E}\{\sup_{A\in\mathcal{A}}|\mu_n(A)-\mu(A)|\} = O(1/\sqrt{n})$ whenever the integral in the bound is uniformly bounded over all $x_1,\ldots,x_n$ and all n. Note that the bound of Theorem 1 is always of larger order of magnitude, trivial cases excepted. The main additional idea is Dudley's *chaining* trick.

PROOF. As in the proof of Theorem 1, we see that

$$\begin{aligned}
&\mathbf{E}\left\{\sup_{A\in\mathcal{A}}|\mu_n(A)-\mu(A)|\right\}\\
&\le \frac{1}{n}\mathbf{E}\left\{\sup_{A\in\mathcal{A}}\left|\sum_{i=1}^n \sigma_i(\mathbf{1}_{[X_i\in A]}-\mathbf{1}_{[X_i'\in A]})\right|\right\}\\
&\le \frac{1}{n}\mathbf{E}\left\{\sup_{A\in\mathcal{A}}\left|\sum_{i=1}^n \sigma_i\mathbf{1}_{[X_i\in A]}\right|\right\}+\frac{1}{n}\mathbf{E}\left\{\sup_{A\in\mathcal{A}}\left|\sum_{i=1}^n \sigma_i\mathbf{1}_{[X_i'\in A]}\right|\right\}\\
&= \frac{2}{n}\mathbf{E}\left\{\sup_{A\in\mathcal{A}}\left|\sum_{i=1}^n \sigma_i\mathbf{1}_{[X_i\in A]}\right|\right\}\\
&= \frac{2}{n}\mathbf{E}\left\{\mathbf{E}\left\{\sup_{A\in\mathcal{A}}\left|\sum_{i=1}^n \sigma_i\mathbf{1}_{[X_i\in A]}\right|\,\middle|X_1,\ldots,X_n\right\}\right\}.
\end{aligned}$$

Just as in the proof of Theorem 1, we fix the values $X_1=x_1,\ldots,X_n=x_n$ and study

$$\mathbf{E}\left\{\sup_{A\in\mathcal{A}}\left|\sum_{i=1}^n \sigma_i\mathbf{1}_{[x_i\in A]}\right|\right\}=\mathbf{E}\left\{\max_{b\in\mathcal{A}(x_1^n)}\left|\sum_{i=1}^n\sigma_i b_i\right|\right\}.$$

Now let $B_0 \stackrel{\text{def}}{=} \{b^{(0)}\}$ be the singleton set containing the all-zero vector $b^{(0)}=(0,\ldots,0)$, and let $B_1,B_2,\ldots,B_M$ be subsets of $\{0,1\}^n$ such that each B_k is a minimal cover of $\mathcal{A}(x_1^n)$ of radius 2^{-k}, and $M=\lfloor\log_2\sqrt{n}\rfloor+1$. Note that B_0 is also a cover of radius 2^0, and that $B_M=\mathcal{A}(x_1^n)$. Now denote the (random) vector reaching the maximum by $b^*=(b_1^*,\ldots,b_n^*)\in\mathcal{A}(x_1^n)$, that is,

$$\left|\sum_{i=1}^n\sigma_i b_i^*\right| = \max_{b\in\mathcal{A}(x_1^n)}\left|\sum_{i=1}^n\sigma_i b_i\right|,$$

and, for each $k\le M$, let $b^{(k)}\in B_k$ be a nearest neighbor of b^* in the kth cover, that is,

$$\rho(b^{(k)},b^*)\le\rho(b,b^*)\quad\text{for all } b\in B_k.$$

Note that $\rho(b^{(k)},b^*)\le 2^{-k}$, and therefore

$$\rho(b^{(k)},b^{(k-1)})\le\rho(b^{(k)},b^*)+\rho(b^{(k-1)},b^*)\le 3\cdot 2^{-k}.$$

Now clearly,

$$\sum_{i=1}^{n} \sigma_i b_i^* = \sum_{i=1}^{n} \sigma_i b_i^{(0)} + \sum_{k=1}^{M} \sum_{i=1}^{n} \sigma_i (b_i^{(k)} - b_i^{(k-1)})$$
$$= \sum_{k=1}^{M} \sum_{i=1}^{n} \sigma_i (b_i^{(k)} - b_i^{(k-1)}),$$

so

$$\mathbf{E}\left\{ \max_{b \in \mathcal{A}(x_1^n)} \left| \sum_{i=1}^{n} \sigma_i b_i \right| \right\}$$
$$= \mathbf{E}\left\{ \left| \sum_{k=1}^{M} \sum_{i=1}^{n} \sigma_i (b_i^{(k)} - b_i^{(k-1)}) \right| \right\}$$
$$\leq \sum_{k=1}^{M} \mathbf{E}\left\{ \left| \sum_{i=1}^{n} \sigma_i (b_i^{(k)} - b_i^{(k-1)}) \right| \right\}$$
$$\leq \sum_{k=1}^{M} \mathbf{E}\left\{ \max_{b \in B_k, c \in B_{k-1}: \rho(b,c) \leq 3 \cdot 2^{-k}} \left| \sum_{i=1}^{n} \sigma_i (b_i - c_i) \right| \right\}.$$

Now it follows from Hoeffding's Lemma 2.1 that for each pair $b \in B_k, c \in B_{k-1}$ with $\rho(b, c) \leq 3 \cdot 2^{-k}$, and for all $s > 0$,

$$e^{s \sum_{i=1}^{n} \sigma_i (b_i - c_i)} \leq e^{s^2 n \left(3 \cdot 2^{-k}\right)^2 / 2}.$$

On the other hand, the number of such pairs is bounded by $|B_k| \cdot |B_{k-1}| \leq |B_k|^2 = N(2^{-k}, \mathcal{A}(x_1^n))^2$. Then Lemma 2.2 implies that for each $1 \leq k \leq M$:

$$\mathbf{E}\left\{ \max_{b \in B_k, c \in B_{k-1}: \rho(b,c) \leq 3 \cdot 2^{-k}} \left| \sum_{i=1}^{n} \sigma_i \left(b_i - c_i\right) \right| \right\}$$
$$\leq 3\sqrt{n} 2^{-k} \sqrt{2 \log 2N(2^{-k}, \mathcal{A}(x_1^n))^2}.$$

Summarizing, we obtain

$$\mathbf{E}\left\{ \max_{b \in \mathcal{A}(x_1^n)} \left| \sum_{i=1}^{n} \sigma_i b_i \right| \right\} \leq 3\sqrt{n} \sum_{k=1}^{M} 2^{-k} \sqrt{2 \log 2N(2^{-k}, \mathcal{A}(x_1^n))^2}$$
$$\leq 12\sqrt{n} \sum_{k=1}^{\infty} 2^{-(k+1)} \sqrt{\log 2N(2^{-k}, \mathcal{A}(x_1^n))}$$
$$\leq 12\sqrt{n} \int_0^1 \sqrt{\log 2N(r, \mathcal{A}(x_1^n))}\, dr,$$

where at the last step we used the fact that $N(r, \mathcal{A}(x_1^n))$ is a monotonically decreasing function of r. The proof is finished. □

§3.3. Example: The Dvoretzky–Kiefer–Wolfowitz Theorem

An important special case is when $\mathcal{A}$ contains all sets of form $A = (-\infty, x]$, $x \in \mathbf{R}$. In this case, $\mu(A)$ is just the distribution function

$$F(x) = \mathbf{P}\{X \le x\}$$

and $\mu_n(A)$ is the corresponding empirical distribution function

$$F_n(x) = \frac{1}{n}\sum_{i=1}^{n} \mathbf{1}_{[X_i \le x]}.$$

The study of $\sup_x |F(x) - F_n(x)| = \sup_{A\in\mathcal{A}} |\mu(A) - \mu_n(A)|$ is a classical theme of mathematical statistics. Theorem 1 may be applied to obtain

$$\mathbf{E}\left\{\sup_x |F(x) - F_n(x)|\right\} \le 2\sqrt{\frac{\log(2n+2)}{n}}.$$

Theorem 2 implies the following improvement:

THEOREM 3.3 (DVORETZKY–KIEFER–WOLFOWITZ THEOREM). *For every n and for all distributions,*

$$\mathbf{E}\left\{\sup_x |F(x) - F_n(x)|\right\} \le \frac{c}{\sqrt{n}},$$

where c is a universal constant.

From the proof below we see that $c \le 24\sqrt{2\pi}$. This constant is not the smallest possible. The best constant was obtained by Massart (1990), who proved that for every $t > 0$:

$$\mathbf{P}\left\{\sup_x |F(x) - F_n(x)| > t\right\} \le 2e^{-2nt^2}.$$

Recall that the upper bound of Theorem 3 may also be converted into an exponential tail inequality with the help of the bounded difference inequality of the previous chapter. Denoting $Z = \sup_x |F(x) - F_n(x)|$, and using Massart's inequality, we see that

$$\mathbf{E}Z \le \sqrt{\mathbf{E}Z^2} = \sqrt{\int_0^\infty \mathbf{P}\{Z > \sqrt{t}\}\, dt} \le \sqrt{\int_0^\infty 2e^{-2nt}\, dt} = \frac{1}{\sqrt{n}},$$

so, in fact, we may take $c = 1$ in Theorem 3.

PROOF. Fix $x_1, \dots, x_n \in \mathbf{R}$. To apply Theorem 2, we find a good upper bound for $N(r, \mathcal{A}(x_1^n))$. Assume without loss of generality that $x_1 \le x_2 \le \cdots \le x_n$. Observe that $|\mathcal{A}(x_1^n)| \le n+1$, and every $b \in \mathcal{A}(x_1^n)$ has the form $(1, 1, \dots, 1, 0, 0, \dots, 0)$. For any $r \in (0,1)$, we construct a cover of

$\mathcal{A}(x_1^n)$ of radius r by considering only those elements of $\mathcal{A}(x_1^n)$ which contain ik ones, where $k = \lfloor nr^2 \rfloor$, and $i = 1, 2, \ldots, \lfloor n/k \rfloor$. Thus, for every $r \in (0, 1)$,

$$N(r, \mathcal{A}(x_1^n)) \leq \left\lfloor \frac{n}{\lfloor nr^2 \rfloor} \right\rfloor \leq \frac{1}{r^2} + 1.$$

Finally,

$$\int_0^1 \sqrt{\log 2N(r, \mathcal{A}(x_1^n))}\, dr \leq \int_0^1 \sqrt{\log \left(\frac{2}{r^2} + 2 \right)}\, dr$$

$$\leq \int_0^1 \sqrt{\log \frac{4}{r^2}}\, dr \leq \sqrt{2\pi},$$

where we used the transformation $r = 2e^{-u/2}$ and used the gamma integral $\int_0^\infty \sqrt{u} e^{-u}\, du = \sqrt{\pi/4}$. Theorem 2 then gives the result. □

§3.4. Bibliographic Remarks

Uniform deviation inequalities have been sudied in empirical process theory, pattern recognition, and computational learning theory. Theorem 1 is a version of the celebrated Vapnik–Chervonenkis inequality (Vapnik and Chervonenkis, 1971). The simple symmetrization trick used in the proof here is due to Giné and Zinn (1984). Theorem 2 is a simple version of Dudley's metric entropy bound (Dudley 1978). Theorem 3 is due to Dvoretzky, Kiefer, and Wolfowitz (1956), and its sharpest form was given by Massart (1990). There are many excellent texts surveying various aspects of empirical process theory, such as Dudley (1984), Pollard (1989, 1990), Ledoux and Talagrand (1991), Giné (1996), Talagrand (1996), van der Vaart and Wellner (1996), and Vapnik (1998).

§3.5. Exercises

Exercise 3.1 Massart's inequality implies that

$$\mathbf{E}\left\{ \sup_x |F(x) - F_n(x)| \right\} \leq 1/\sqrt{n}$$

(Section 3). Can you improve the constant further? HINT: Split the integral optimally, noting that $\mathbf{P}\{Z > \sqrt{t}\} \leq 1$.

Exercise 3.2 Use the bounded difference inequality to show that for any class of sets $\mathcal{A}$, $\sup_{A \in \mathcal{A}} |\mu_n(A) - \mu(A)| \to 0$ in probability if and only if $\sup_{A \in \mathcal{A}} |\mu_n(A) - \mu(A)| \to 0$ with probability one.

Exercise 3.3 (INEQUALITIES FOR THE RELATIVE DEVIATION). Show the following one-sided improvement of Theorem 1: for every $\epsilon > 0$:

$$\mathbf{P}\left\{\sup_{A\in\mathcal{A}} \frac{\mu(A)-\mu_n(A)}{\sqrt{\mu(A)}} > \epsilon\right\} \le 4\mathbb{S}_{\mathcal{A}}(2n)e^{-n\epsilon^2/4}$$

(Vapnik and Chervonenkis, 1974; see also Anthony and Shawe-Taylor, 1993). HINT: Using the notation of the proof of Theorem 1, proceed as follows:

A.

$$\mathbf{P}\left\{\sup_{A\in\mathcal{A}} \frac{\mu(A)-\mu_n(A)}{\sqrt{\mu(A)}} > \epsilon\right\} \le 2\mathbf{P}\left\{\sup_{A\in\mathcal{A}} \frac{\mu_n'(A)-\mu_n(A)}{\sqrt{(\mu_n'(A)+\mu_n(A))/2}} > \epsilon\right\}.$$

B.

$$\mathbf{P}\left\{\sup_{A\in\mathcal{A}} \frac{\mu_n'(A)-\mu_n(A)}{\sqrt{(\mu_n'(A)+\mu_n(A))/2}} > \epsilon\right\} =$$

$$\mathbf{E}\left\{\mathbf{P}\left\{\sup_{A\in\mathcal{A}} \frac{(1/n)\sum_{i=1}^n \sigma_i(\mathbb{1}_{X_i'\in A}-\mathbb{1}_{X_i\in A})}{\sqrt{(\mu_n'(A)+\mu_n(A))/2}} > \epsilon \,\middle|\, X_1,X_1',\ldots,X_n,X_n'\right\}\right\}.$$

C. Use the union bound and Hoeffding's inequality to bound the conditional probability inside.

Exercise 3.4 (CONTINUATION). Proceed as in the previous exercise to show

$$\mathbf{P}\left\{\sup_{A\in\mathcal{A}} \frac{\mu_n(A)-\mu(A)}{\sqrt{\mu_n(A)}} > \epsilon\right\} \le 4\mathbb{S}_{\mathcal{A}}(2n)e^{-n\epsilon^2/4}.$$

Exercise 3.5 (CONTINUATION). Deduce from the previous two exercises that for all $t \in (0,1)$ and $s > 0$:

$$\mathbf{P}\left\{\sup_{A\in\mathcal{A}} \frac{\mu(A)-\mu_n(A)}{\mu(A)+\mu_n(A)+s/2} > t\right\} \le 4\mathbb{S}_{\mathcal{A}}(2n)e^{-nst^2/4}$$

and

$$\mathbf{P}\left\{\sup_{A\in\mathcal{A}} \frac{\mu_n(A)-\mu(A)}{\mu(A)+\mu_n(A)+s/2} > t\right\} \le 4\mathbb{S}_{\mathcal{A}}(2n)e^{-nst^2/4}.$$

These inequalities are due to Pollard (1990) and Haussler (1992). The fact that they follow from Exercises 3 and 4 is pointed out by Bartlett and Lugosi (1999). HINT: Take $\alpha > 0$. Considering the cases $\mu(A) < (\alpha+1)^2\epsilon^2\alpha^{-2}$ and $\mu(A) \ge (\alpha+1)^2\epsilon^2\alpha^{-2}$ separately, show that $\mu(A)-\mu_n(A) \le \epsilon\sqrt{\mu(A)}$ implies that $\mu(A) \le (1+\alpha)\mu_n(A) + \epsilon^2(1+\alpha)/\alpha$. Then choose

$\alpha = 2t/(1-t)$ and $\epsilon^2 = st^2/(1-t^2)$ to prove that Exercise 3 implies the first inequality. The second inequality follows similarly from Exercise 4.

Exercise 3.6 Use Exercise 3 to show that for all $t \in (0,1]$ and $\epsilon > 0$:

$$\mathbf{P}\{\exists\, A \in \mathcal{A} : \mu(A) > \epsilon \text{ and } \mu_n(A) \leq (1-t)\mu(A)\} \leq 4\mathbb{S}_{\mathcal{A}}(2n)e^{-n\epsilon t^2/4}.$$

What do you obtain by setting $t = 1$?

§3.6. References

M. Anthony and J. Shawe–Taylor, "A result of Vapnik with applications," *Discrete Applied Mathematics*, vol. 47, pp. 207–217, 1993.

P. Bartlett and G. Lugosi, "An inequality for uniform deviations of sample averages from their means," *Statistics and Probability Letters*, vol. 44, pp. 55–62, 1999.

R. M. Dudley, "Central limit theorems for empirical measures," *Annals of Probability*, vol. 6, pp. 899–929, 1978.

R. M. Dudley, "Empirical Processes," in: *Ecole de Probabilité de St. Flour 1982*, Springer-Verlag, New York, 1984.

A. Dvoretzky, J. Kiefer, and J. Wolfowitz, "Asymptotic minimax character of a sample distribution function and of the classical multinomial estimator," *Annals of Mathematical Statistics*, vol. 33, pp. 642–669, 1956.

E. Giné, "Empirical processes and applications: An overview," *Bernoulli*, vol. 2, pp. 1–28, 1996.

E. Giné and J. Zinn, "Some limit theorems for empirical processes," *Annals of Probability*, vol. 12, pp. 929–989, 1984.

D. Haussler, "Decision theoretic generalizations of the PAC model for neural net and other learning applications," *Information and Computation*, vol. 100, pp. 78–150, 1992.

M. Ledoux and M. Talagrand, *Probability in Banach Space*, Springer-Verlag, New York, 1991.

P. Massart, "The tight constant in the Dvoretzky–Kiefer–Wolfowitz inequality," *Annals of Probability*, vol. 18, pp. 1269–1283, 1990.

D. Pollard, "Asymptotics via empirical processes," *Statistical Science*, vol. 4, pp. 341–366, 1989.

D. Pollard, *Empirical Processes: Theory and Applications*, NSF-CBMS Regional Conference Series in Probability and Statistics, Institute of Mathematical Statistics, Hayward, CA, 1990.

M. Talagrand, "Majorizing measures: the generic chaining," *Annals of Probability*, vol. 24, pp. 1049–1103, 1996.

A. W. van der Vaart and J. A. Wellner, *Weak Convergence and Empirical Processes*, Springer-Verlag, New York, 1996.

V. N. Vapnik, *Statistical Learning Theory*, Wiley, New York, 1998.

V. N. Vapnik and A. Ya. Chervonenkis, "On the uniform convergence of relative frequencies of events to their probabilities," *Theory of Probability and its Applications*, vol. 16, pp. 264–280, 1971.

V. N. Vapnik and A. Ya. Chervonenkis, *Theory of Pattern Recognition*, Nauka, Moscow, 1974.

Chapter 4
Combinatorial Tools

§4.1. Shatter Coefficients

Consider a class $\mathcal{A}$ of subsets of $\mathbf{R}^d$, and let $x_1, \ldots, x_n \in \mathbf{R}^d$ be arbitrary points. Recall from the previous chapter that properties of the finite set $\mathcal{A}(x_1^n) \subset \{0,1\}^n$ defined by

$$\mathcal{A}(x_1^n) = \{b = (b_1, \ldots, b_n) \in \{0,1\}^n : \exists\, A \in \mathcal{A} : b_i = \mathbf{1}_{[x_i \in A]},\ i = 1, \ldots, n\}$$

play an essential role in bounding uniform deviations of the empirical measure. In particular, the maximal cardinality of $\mathcal{A}(x_1^n)$:

$$\mathrm{S}_{\mathcal{A}}(n) = \max_{x_1, \ldots, x_n \in \mathbf{R}^d} |\mathcal{A}(x_1^n)|$$

(i.e., the shatter coefficient) provides simple bounds via Theorem 3.1. We have also seen that more detailed knowledge about the covering numbers of $\mathcal{A}(x_1^n)$ may lead to sharper bounds. This chapter is devoted to various combinatorial techniques to bound these quantities. We begin with some elementary properties of the shatter coefficient.

THEOREM 4.1. *Let $\mathcal{A}$ and $\mathcal{B}$ be classes of subsets of $\mathbf{R}^d$, and let $n, m \geq 1$ be integers. Then:*

(1) $\mathrm{S}_{\mathcal{A}}(n+m) \leq \mathrm{S}_{\mathcal{A}}(n)\mathrm{S}_{\mathcal{A}}(m)$;
(2) *if* $\mathcal{C} = \mathcal{A} \cup \mathcal{B}$, *then* $\mathrm{S}_{\mathcal{C}}(n) \leq \mathrm{S}_{\mathcal{A}}(n) + \mathrm{S}_{\mathcal{B}}(n)$;
(3) *if* $\mathcal{C} = \{C = A^c : A \in \mathcal{A}\}$, *then* $\mathrm{S}_{\mathcal{C}}(n) = \mathrm{S}_{\mathcal{A}}(n)$;
(4) *if* $\mathcal{C} = \{C = A \cap B : A \in \mathcal{A} \text{ and } B \in \mathcal{B}\}$, *then* $\mathrm{S}_{\mathcal{C}}(n) \leq \mathrm{S}_{\mathcal{A}}(n)\mathrm{S}_{\mathcal{B}}(n)$;
(5) *if* $\mathcal{C} = \{C = A \cup B : A \in \mathcal{A} \text{ and } B \in \mathcal{B}\}$, *then* $\mathrm{S}_{\mathcal{C}}(n) \leq \mathrm{S}_{\mathcal{A}}(n)\mathrm{S}_{\mathcal{B}}(n)$;
(6) *if* $\mathcal{C} = \{C = A \times B : A \in \mathcal{A} \text{ and } B \in \mathcal{B}\}$, *then* $\mathrm{S}_{\mathcal{C}}(n) \leq \mathrm{S}_{\mathcal{A}}(n)\mathrm{S}_{\mathcal{B}}(n)$.

PROOF. Parts (1), (2), (3), and (6) are immediate from the definition. To show (4), fix $x_1, \ldots, x_n$, let $N = |\mathcal{A}(x_1^n)| \leq \mathrm{S}_{\mathcal{A}}(n)$, and denote by $A_1, A_2, \ldots, A_N$ the different sets of the form $\{x_1, \ldots, x_n\} \cap A$ for some $A \in \mathcal{A}$. For all $1 \leq i \leq N$, sets in $\mathcal{B}$ pick at most $\mathrm{S}_{\mathcal{B}}(|A_i|) \leq \mathrm{S}_{\mathcal{B}}(n)$ different subsets of A_i. Thus,

$$|\mathcal{A}(x_1^n)| \leq \sum_{i-1}^{N} \mathrm{S}_{\mathcal{B}}(|A_i|) \leq \mathrm{S}_{\mathcal{A}}(n)\mathrm{S}_{\mathcal{B}}(n).$$

Part (5) follows from (4) and (3). □

§4.2. Vapnik–Chervonenkis Dimension and Shatter Coefficients

The *Vapnik–Chervonenkis dimension* V of a class $\mathcal{A}$ of sets is defined as the largest integer n such that

$$\mathrm{S}_{\mathcal{A}}(n) = 2^n.$$

If $\mathrm{S}_{\mathcal{A}}(n) = 2^n$ for all n, then we say that $V = \infty$. Clearly, if $\mathrm{S}_{\mathcal{A}}(n) < 2^n$ for some n, then for all $m > n$, $\mathrm{S}_{\mathcal{A}}(m) < 2^m$, and therefore the Vapnik–Chervonenkis dimension is always well defined. As the next basic result shows, the Vapnik–Chervonenkis dimension provides a useful bound for the shatter coefficient of a class.

THEOREM 4.2 (SAUER'S LEMMA). *Let $\mathcal{A}$ be a class of sets with Vapnik–Chervonenkis dimension $V < \infty$. Then for all n,*

$$\mathrm{S}_{\mathcal{A}}(n) \le \sum_{i=0}^{V} \binom{n}{i}.$$

PROOF. Fix $x_1, \dots, x_n$, such that $|\mathcal{A}(x_1^n)| = \mathrm{S}_{\mathcal{A}}(n)$. Denote $B_0 = \mathcal{A}(x_1^n) \in \{0,1\}^n$. We say that a set $B \subset \{0,1\}^n$ *shatters* a set $S = \{s_1, \dots, s_m\} \subset \{1, 2, \dots, n\}$ if the restriction of B to the components $s_1, \dots, s_m$ is the full m-dimensional binary hypercube, that is,

$$\{(b_{s_1}, \dots, b_{s_m}) : b = (b_1, \dots, b_n) \in B\} = \{0,1\}^m.$$

It suffices to show that the cardinality of any set $B_0 \subset \{0,1\}^n$ that cannot shatter any set of size $m > V$, is at most $\sum_{i=0}^{V} \binom{n}{i}$. This is done by transforming B_0 into a set B_n with $|B_n| = |B_0|$ such that any set shattered by B_n is also shattered by B_0. Moreover, it will be easy to see that $|B_n| \le \sum_{i=0}^{V} \binom{n}{i}$.

For every vector $b = (b_1, \dots, b_n) \in B_0$, if $b_1 = 1$, then flip the first component of b to zero unless $(0, b_2, \dots, b_n) \in B_0$. If $b_1 = 0$, then keep the vector unchanged. The set of vectors B_1 obtained this way obviously has the same cardinality as that of B_0. Moreover, if B_1 shatters a set $S = \{s_1, s_2, \dots, s_m\} \subset \{1, \dots, n\}$, then B_0 also shatters S. This is trivial if $1 \notin S$. If $1 \in S$, then we may assume without loss of generality that $s_1 = 1$. The fact that B_1 shatters S implies that for any $v \in \{0,1\}^{m-1}$ there exists a $b \in B_1$ such that $b_1 = 1$ and $(b_{s_2}, \dots, b_{s_m}) = v$. By the construction of B_1 this is only possible if for any $u \in \{0,1\}^m$ there exists a $b' \in B_0$ such that $(b'_{s_1}, \dots, b'_{s_m}) = u$. This means that B_0 also shatters S.

Now starting from B_1, execute the same transformation, but now by flipping the second component of each vector, if necessary. Again, the cardinality of the obtained set B_2 remains unchanged, and any set shattered by B_2 is also shattered by B_1 (and therefore also by B_0). Repeat the transformation for all components, arriving at the set B_n. Clearly, B_n cannot shatter sets of cardinality larger than V, since otherwise B_0 would shatter sets of the same size. On the other hand, it is easy to see that B_n is such that for every $b \in B_n$, all vectors of form $c = (c_1, \ldots, c_n)$ with $c_i \in \{b_i, 0\}$ for $1 \leq i \leq n$ are also in B_n. Then B_n is a subset of a set of form

$$T = \{b \in \{0,1\}^n : b_i = 0 \text{ if } v_i = 0\},$$

where $v = (v_1, \ldots, v_n)$ is a fixed vector containing at most V ones. This implies that

$$\mathbb{S}_{\mathcal{A}}(n) = |B_0| = |B_n| \leq |T| = \sum_{i=0}^{V} \binom{n}{i},$$

concluding the proof. □

The following corollary makes the meaning of Sauer's lemma more transparent:

COROLLARY 4.1. *Let $\mathcal{A}$ be a class of sets with Vapnik–Chervonenkis dimension $V < \infty$. Then for all n:*

$$\mathbb{S}_{\mathcal{A}}(n) \leq (n+1)^V,$$

and for all $n \geq V$:

$$\mathbb{S}_{\mathcal{A}}(n) \leq \left(\frac{ne}{V}\right)^V.$$

PROOF. By the binomial theorem,

$$(n+1)^V = \sum_{i=0}^{V} n^i \binom{V}{i} = \sum_{i=0}^{V} \frac{n^i V!}{i!(V-i)!} \geq \sum_{i=0}^{V} \frac{n^i}{i!} \geq \sum_{i=0}^{V} \binom{n}{i}.$$

On the other hand, if $V/n \leq 1$, then

$$\left(\frac{V}{n}\right)^V \sum_{i=0}^{V} \binom{n}{i} \leq \sum_{i=0}^{V} \left(\frac{V}{n}\right)^i \binom{n}{i} \leq \sum_{i=0}^{n} \left(\frac{V}{n}\right)^i \binom{n}{i} = \left(1 + \frac{V}{n}\right)^n \leq e^V,$$

where again we used the binomial theorem. □

Recalling Theorem 3.1, we see that if $\mathcal{A}$ is any class of sets with Vapnik–Chervonenkis dimension V, then

$$\mathbf{E}\left\{\sup_{A \in \mathcal{A}} |\mu_n(A) - \mu(A)|\right\} \leq 2\sqrt{\frac{V \log(n+1) + \log 2}{n}},$$

that is, whenever $\mathcal{A}$ has a finite Vapnik–Chervonenkis dimension, the expected largest deviation over $\mathcal{A}$ converges to zero at a rate $O(\sqrt{\log n / n})$. This rate may be improved with the help of a related combinatorial result shown in the next section.

§4.3. Vapnik–Chervonenkis Dimension and Covering Numbers

In the theorem below we relate the Vapnik–Chervonenkis dimension of a class of sets $\mathcal{A}$ to the covering numbers $N(r, \mathcal{A}(x_1^n))$ appearing in Theorem 3.2.

THEOREM 4.3. *Let $\mathcal{A}$ be a class of sets with Vapnik–Chervonenkis dimension $V < \infty$. For every $x_1, \ldots, x_n \in \mathbf{R}^d$ and $0 \leq r \leq 1$,*

$$N(r, \mathcal{A}(x_1^n)) \leq \left(\frac{4e}{r^2}\right)^{V/(1-1/e)}.$$

Theorem 3 is due to Dudley (1978). Haussler (1995) refined Dudley's probabilistic argument and showed that the stronger bound

$$N(r, \mathcal{A}(x_1^n)) \leq e(V+1)\left(\frac{2e}{r^2}\right)^{V}.$$

also holds.

PROOF. Fix $x_1, \ldots, x_n$, and consider the set $B_0 = \mathcal{A}(x_1^n) \in \{0,1\}^n$. Fix $r \in (0,1)$, and let $B_r \subset \{0,1\}^n$ be a minimal cover of B_0 of radius r with respect to the metric

$$\rho(b,c) = \sqrt{\frac{1}{n}\sum_{i=1}^{n} \mathbf{1}_{[b_i \neq c_i]}}.$$

We need to show that $|B_r| \leq \left(4e/r^2\right)^{V/(1-1/e)}$.

First note that there exists a "packing set" $C_r \subset B_0$ such that $|B_r| \leq |C_r|$ and any two elements $b, c \in C_r$ are *r-separated*, that is, $\rho(b,c) > r$. To see this, suppose that C_r is such an r-separated set of maximal cardinality. Then for any $b \in B_0$, there exists a $c \in C_r$ with $\rho(b,c) \leq r$, since otherwise adding b to the set C_r would increase its cardinality, and it would still be r-separated. Thus, C_r is a cover of radius r, which implies that $|B_r| \leq |C_r|$. Denote the elements of C_r by $c^{(1)}, \ldots, c^{(M)}$, where $M = |C_r|$. For any $i, j \leq M$, define $A_{i,j}$ as the set of indices where the binary vectors $c^{(i)}$ and $c^{(j)}$ disagree:

$$A_{i,j} = \{1 \leq m \leq n : c_m^{(i)} \neq c_m^{(j)}\}.$$

Note that any two elements of C_r differ in at least nr^2 components. Next define K independent random variables $Y_1, \ldots, Y_K$, distributed uniformly over the set $\{1, 2, \ldots, n\}$, where K will be specified later. Then for any $i, j \leq M$, $i \neq j$, and $k \leq K$,

$$\mathbf{P}\{Y_k \in A_{i,j}\} \geq r^2,$$

and therefore the probability that no one of $Y_1, \ldots, Y_K$ falls in the set $A_{i,j}$ is less than $\left(1 - r^2\right)^K$. Observing that there are less than M^2 sets $A_{i,j}$, and applying the union bound, we obtain that

$$\mathbf{P}\{\text{for all } i \neq j, i, j \leq M, \text{at least one } Y_k \text{ falls in } A_{i,j}\}$$
$$\geq 1 - M^2(1-r^2)^K \geq 1 - M^2 e^{-Kr^2}.$$

If we choose $K = \lceil 2\log M/r^2 \rceil + 1$, then the above probability is strictly positive. This implies that there exist $K = \lceil 2\log M/r^2 \rceil + 1$ indices $y_1, \ldots, y_K \in \{1, 2, \ldots, n\}$ such that at least one y_k falls in each set $A_{i,j}$. Therefore, restricted to the K components $y_1, \ldots, y_K$, the elements of C_r are all different, and since $C_r \subset B_0$, C_r does not shatter any set of size larger than V. Therefore, by Sauer's lemma we obtain

$$|C_r| = M \leq \left(\frac{eK}{V}\right)^V$$

for $K \leq V$. Thus, if $\log M \geq V$, then

$$\begin{aligned}
\log M &\leq V \log \frac{e\left(\lceil 2\log M/r^2 \rceil + 1\right)}{V} \\
&\leq V\left(\log \frac{4e}{r^2} + \log \frac{\log M}{V}\right) \\
&\leq V \log \frac{4e}{r^2} + \frac{1}{e}\log M \quad (\text{since } \log x < x/e \text{ for } x > 0).
\end{aligned}$$

Therefore,

$$\log M \leq \frac{V}{1 - 1/e} \log \frac{4e}{r^2}.$$

If $\log M < V$, then the above inequality holds trivially. This concludes the proof. □

Combining this result with Theorem 3.2 we obtain that for any class $\mathcal{A}$ with Vapnik–Chervonenkis dimension V:

$$\mathbf{E}\left\{\sup_{A \in \mathcal{A}} |\mu_n(A) - \mu(A)|\right\} \leq c\sqrt{\frac{V}{n}},$$

where c is a universal constant.

§4.4. Examples

Here we calculate the Vapnik–Chervonenkis dimension of some simple classes, which will turn out to be useful later.

LEMMA 4.1. *If $\mathcal{A}$ is the class of all rectangles in $\mathbf{R}^d$, then $V = 2d$.*

PROOF. To see that there are $2d$ points that can be shattered by $\mathcal{A}$, just consider the $2d$ vectors with $d-1$ zero components, and one nonzero component which is either 1 or -1. On the other hand, for any given set of $2d+1$ points we can choose a subset of at most $2d$ points with the property that it contains a point with largest first coordinate, a point with smallest first coordinate, a point with largest second coordinate, and so forth. Clearly, there is no set in $\mathcal{A}$ which contains these points, but not the rest. □

LEMMA 4.2. *Let $\mathcal{G}$ be an m-dimensional vector space of real-valued functions defined on $\mathbf{R}^d$. The class of sets*

$$\mathcal{A} = \{\{x : g(x) \geq 0\} : g \in \mathcal{G}\}$$

has Vapnik–Chervonenkis dimension $V \leq m$.

PROOF. It suffices to show that no set of size $m+1$ can be shattered by sets of the form $\{x : g(x) \geq 0\}$. Fix $m+1$ arbitrary points $x_1, \ldots, x_{m+1}$, and define the linear mapping $L : \mathcal{G} \to \mathbf{R}^{m+1}$ as

$$L(g) = (g(x_1), \ldots, g(x_{m+1})).$$

Then the image of $\mathcal{G}$, $L(\mathcal{G})$, is a linear subspace of $\mathbf{R}^{m+1}$ of dimension not exceeding m. This implies the existence of a nonzero vector $\gamma = (\gamma_1, \ldots, \gamma_{m+1}) \in \mathbf{R}^{m+1}$ orthogonal to $L(\mathcal{G})$, that is, for every $g \in \mathcal{G}$,

$$\gamma_1 g(x_1) + \ldots + \gamma_{m+1} g(x_{m+1}) = 0.$$

We may assume that at least one of the γ_i's is negative. Rearranging this equality so that all terms with nonnegative γ_i stay on the left-hand side, we get

$$\sum_{i:\gamma_i \geq 0} \gamma_i g(x_i) = \sum_{i:\gamma_i < 0} -\gamma_i g(x_i).$$

Now suppose that there exists a $g \in \mathcal{G}$ such that the set $\{x : g(x) \geq 0\}$ picks exactly the x_i's on the left-hand side. Then all terms on the left-hand side are nonnegative, while the terms on the right-hand side must be negative, which is a contradiction, so $x_1, \ldots, x_{m+1}$ cannot be shattered, which implies the statement. □

Generalizing a result of Schläffli (1950), Cover (1965) showed that if $\mathcal{G}$ is defined as the linear space of functions spanned by functions $\psi_1, \ldots, \psi_m : \mathbf{R}^d \to \mathbf{R}$, and the vectors $\Psi(x_i) = (\psi_1(x_i), \ldots, \psi_m(x_i))$, $i = 1, 2, \ldots, n$, are linearly independent, then for the class of sets $\mathcal{A} = \{\{x : g(x) \geq 0\} : g \in \mathcal{G}\}$ we have

$$|\mathcal{A}(x_1^n)| = 2 \sum_{i=0}^{m-1} \binom{n-1}{i},$$

which often gives a slightly sharper estimate than Lemma 2. The proof is left as an exercise. From Lemma 2 we immediately deduce the following:

COROLLARY 4.2.

(1) *If $\mathcal{A}$ is the class of all linear half-spaces, that is, subsets of $\mathbf{R}^d$ of the form $\{x : a^T x \geq b\}$, where $a \in \mathbf{R}^d, b \in \mathbf{R}$, take all possible values, then $V \leq d+1$.*

(2) *If $\mathcal{A}$ is the class of all closed balls in $\mathbf{R}^d$, that is, sets of the form*

$$\left\{ x = (x^{(1)}, \ldots, x^{(d)}) : \sum_{i=1}^{d} |x^{(i)} - a_i|^2 \leq b \right\}, \quad a_1, \ldots, a_d, \ b \in \mathbf{R},$$

then $V \leq d+2$.

(3) *If $\mathcal{A}$ is the class of all ellipsoids in $\mathbf{R}^d$, that is, sets of the form $\{x : x^T \Sigma^{-1} x \leq 1\}$, where Σ is a positive definite symmetric matrix, then $V \leq d(d+1)/2+1$.*

Note that the above-mentioned result implies that the Vapnik–Chervonenkis dimension of the class of all linear half-spaces actually equals $d+1$. Dudley (1979) proved that in the case of the class of all closed balls the above inequality is not tight, and the Vapnik–Chervonenkis dimension equals $d+1$ (see Exercise 5).

§4.5. Bibliographic Remarks

Sauer's lemma was independently proved by Vapnik and Chervonenkis (1971), Sauer (1972), and Shelah (1972). The proof presented here is due to Steele (1978), see also Frankl (1983). Various extensions and generalizations may be found in Szarek and Talagrand (1997), Alesker (1997), Alon, Ben-David, Cesa-Bianchi and Haussler (1997), Cesa-Bianchi and Haussler (1998). For a comprehensive treatment of related issues we refer to the book of Anthony and Bartlett (1999). Theorem 3 was first proved by Dudley (1978). The strongest known form was obtained by Haussler (1995). Lemma 2 appears in Steele (1975) and Dudley (1978).

§4.6. Exercises

Exercise 4.1 Let $\mathcal{A}$ and $\mathcal{B}$ be classes of sets with Vapnik–Chervonenkis dimension V_1 and V_2, respectively. Show that the Vapnik–Chervonenkis dimension of $\mathcal{A} \cup \mathcal{B}$ is not more than $V_1 + V_2 + 1$, and show an example when equality is achieved (Assouad, 1981). HINT: Use Sauer's lemma with $n = V_1 + V_2 + 2$.

Exercise 4.2 Let $\mathcal{A}$ be a class of sets with Vapnik–Chervonenkis dimension V, and consider the class of sets $\mathcal{B}$ obtainable by any set-theoretic formula of k variables (using $\cap, \cup$, and set difference) whose variables are sets in $\mathcal{A}$. Show that the Vapnik–Chervonenkis dimension of $\mathcal{B}$ is at most $2Vk\log_2(2Vk)$.

Exercise 4.3 Prove by induction that if $\mathcal{G}$ is the linear space of functions spanned by functions $\psi_1, \ldots, \psi_m : \mathbf{R}^d \to \mathbf{R}$, and the vectors $\Psi(x_i) = (\psi_1(x_i), \ldots, \psi_m(x_i))$, $i = 1, 2, \ldots, n$ are linearly independent, then for the class of sets $\mathcal{A} = \{\{x : g(x) \geq 0\} : g \in \mathcal{G}\}$ we have

$$|\mathcal{A}(x_1^n)| = 2\sum_{i=0}^{m-1}\binom{n-1}{i}.$$

Deduce that the Vapnik–Chervonenkis dimension of $\mathcal{A}$ is m (Schläffli, 1950; Cover, 1965).

Exercise 4.4 Let $\mathcal{G}$ be an m-dimensional vector space of functions defined on $\mathbf{R}^d$, and let $f \notin \mathcal{G}$ be a fixed function. Prove that the class of sets

$$\mathcal{A} = \{\{x : f(x) + g(x) > 0\} : g \in \mathcal{G}\}$$

has Vapnik–Chervonenkis dimension m (Wenocur and Dudley, 1981).

Exercise 4.5 Use the previous exercise to deduce that the Vapnik–Chervonenkis dimension of the class of all closed balls in $\mathbf{R}^d$ is $d+1$ (Dudley, 1979). HINT: Take $f(x) = -\|x\|^2$.

Exercise 4.6 (CLASSES OBTAINED BY SCALING). Let $A \subset \mathbf{R}^d$ be a nonempty star-shaped set (a set A is star-shaped if $x \in A$ implies that $\lambda x \in A$ for all $\lambda \in [0, 1]$). For any $r > 0$, define the set $r \cdot A = \{x \in \mathbf{R}^d : x/r \in A\}$, and consider the one-parameter family of sets $\mathcal{A} = \{r \cdot A : r > 0\}$. Show that the Vapnik–Chervonenkis dimension of $\mathcal{A}$ is one.

Exercise 4.7 (CONTINUATION). Construct a set $A \subset \mathbf{R}$ such that the Vapnik–Chervonenkis dimension of $\mathcal{A} = \{r \cdot A : r > 0\}$ is infinite.

Exercise 4.8 (TRANSLATION CLASSES). For a fixed set A in $\mathbf{R}^d$, consider the class of sets of all translates $x + A$ of A. Give an example for A such that the Vapnik–Chervonenkis dimension of the obtained class is infinite. Investigate what kind of conditions for A are sufficient for the finiteness of the Vapnik–Chervonenkis dimension. (Note: Exercise 4 above might help in the last part.)

Exercise 4.9 (CONVEX SETS). Show that the Vapnik–Chervonenkis dimension of the class $\mathcal{A}$ of all convex sets in $\mathbf{R}^d$ is infinite. Construct a distribution such that

$$\mathbf{E}\left\{\sup_{A\in\mathcal{A}} |\mu_n(A) - \mu(A)|\right\}$$

does not converge to zero as $n \to \infty$.

Exercise 4.10 (CONTINUATION). Assume that the i.i.d. random variables $X_1, \ldots, X_n$ have a common density f in $\mathbf{R}^d$. Prove that for each f,

$$\lim_{n\to\infty} \log \mathbf{E}\{|\mathcal{A}(X_1^n)|\} = 0.$$

Show that this implies

$$\lim_{n\to\infty} \mathbf{E}\left\{\sup_{A\in\mathcal{A}} |\mu_n(A) - \mu(A)|\right\} = 0.$$

Exercise 4.11 (PACKING NUMBERS). For a set $B \subset \{0,1\}^n$, define the *packing number* $M(r, B)$ of radius $r > 0$ as the cardinality of the largest r-separated subset of B. See the proof of Theorem 3 for the definition. Prove the following relation between covering numbers and packing numbers:

$$N(r, B) \le M(r, B) \le N(r/2, B).$$

Exercise 4.12 CONTINUATION. Let V be a fixed positive integer, and assume that $n = mV$ is an integer multiple of V. Show that there exists a set $B \subset \{0,1\}^n$ with Vapnik–Chervonenkis dimension V such that for each $k = 1, \ldots, n$,

$$M(\sqrt{k/n}, B) \ge \left(\frac{n}{2e(k+V)}\right)^V$$

(Haussler, 1995). This shows that Theorem 3 cannot be substantially improved. HINT: Choose B to be the set of all concatenations of V vectors of m components of the form $(00\cdots011\cdots1)$.

§4.7. References

S. Alesker, "A remark on the Szarek–Talagrand theorem," *Combinatorics, Probability, and Computing*, vol. 6, pp. 139–144, 1997.

N. Alon, S. Ben–David, N. Cesa–Bianchi, and D. Haussler, "Scale-sensitive dimensions, uniform convergence, and learnability," *Journal of the ACM*, vol. 44, pp. 615–631, 1997.

M. Anthony and P. L. Bartlett, *Neural Network Learning: Theoretical Foundations*, Cambridge University Press, Cambridge, 1999.

P. Assouad, "Sur les classes de Vapnik–Chervonenkis," *Comptes Rendus des Séances de l'Académie des Sciences. Série I*, vol. 292, pp. 921–924, 1981.

N. Cesa–Bianchi and D. Haussler, "A graph-theoretic generalization of the Sauer-Shelah lemma," *Discrete Applied Mathematics*, vol. 86, pp. 27–35, 1998.

T. M. Cover, "Geometrical and statistical properties of systems of linear inequalities with applications in pattern recognition," *IEEE Transactions on Electronic Computers*, vol. 14, pp. 326–334, 1965.

R. M. Dudley, "Central limit theorems for empirical measures," *Annals of Probability*, vol. 6, pp. 899–929, 1978.

R. M. Dudley, "Balls in R^k do not cut all subsets of $k+2$ points," *Advances in Mathematics*, vol. 31 (3), pp. 306–308, 1979.

P. Frankl, "On the trace of finite sets," *Journal of Combinatorial Theory, Series A*, vol. 34, pp. 41–45, 1983.

D. Haussler, "Sphere packing numbers for subsets of the Boolean n-cube with bounded Vapnik-Chervonenkis dimension," *Journal of Combinatorial Theory, Series A*, vol. 69, pp. 217–232, 1995.

N. Sauer, "On the density of families of sets," *Journal of Combinatorial Theory Series A*, vol. 13, pp. 145–147, 1972.

L. Schläffli, *Gesammelte Mathematische Abhandlungen*, Birkhäuser-Verlag, Basel, 1950.

S. Shelah, "A combinatorial problem: Stability and order for models and theories in infinity languages," *Pacific Journal of Mathematics*, vol. 41, pp. 247–261, 1972.

J. M. Steele, "Combinatorial entropy and uniform limit laws," PhD dissertation, Stanford University, Stanford, CA, 1975.

J. M. Steele, "Existence of submatrices with all possible columns," *Journal of Combinatorial Theory, Series A*, vol. 28, pp. 84–88, 1978.

S. J. Szarek and M. Talagrand, "On the convexified Sauer–Shelah theorem," *Journal of Combinatorial Theory, Series B*, vol. 69, pp. 183–192, 1997.

V. N. Vapnik and A. Ya. Chervonenkis, "On the uniform convergence of relative frequencies of events to their probabilities," *Theory of Probability and its Applications*, vol. 16, pp. 264–280, 1971.

R. S. Wenocur and R. M. Dudley, "Some special Vapnik–Chervonenkis classes," *Discrete Mathematics*, vol. 33, pp. 313–318, 1981.

Chapter 5
Total Variation

§5.1. Density Estimation

A random variable X has a density f on $\mathbf{R}^d$ when for all Borel sets $A \subseteq \mathbf{R}^d$, $\mathbf{P}\{X \in A\} = \int_A f(x)\,dx$. In other words, if A is a small ball about x, its probability is about $f(x)$ times the volume of A. The purpose of density estimation is to estimate an unknown density f from an i.i.d. sample $X_1, \ldots, X_n$ drawn from f. There are other questions one may ask, like the estimation of a functional such as $\int f^\alpha(x)\,dx$ or $\int \psi(f(x), x)\,dx$ for general ψ. These are not dealt with in the text and may indeed require a very different treatment. We are concerned with the density itself as a means of computing probabilities of all sets. A density estimate $f_n(x) = f_n(x; X_1, \ldots, X_n)$ is thus a mapping from $(\mathbf{R}^d)^{n+1}$ to $\mathbf{R}$. Its quality is measured by how well it performs the task at hand, estimating probabilities. In this respect, the total variation criterion is a natural distance:

$$\sup_A \left| \int_A f_n - \int_A f \right| .$$

If this is smaller than ϵ, then all probabilities will be estimated with errors not exceeding ϵ. We realize that there are other possible distance measures between densities, such as the Hellinger distance and the Kullback–Leibler based distances. Some of these are introduced in the exercises, and their relationship with the total variation criterion is explored there. Standard L_p distances, with $p \neq 1$, are not well suited for our problem, and this will become apparent from various examples in the exercises of this and future chapters.

Densities occur also indirectly in many processes, for example, as stationary, implicit or asymptotic distributions. We will not be concerned with such situations, and will not deal with dependent samples and other interesting settings. The remainder of this chapter is meant to familiarize the reader with the total variation criterion and the L_1 distance, and to present some fundamental properties of densities such as the Lebesgue density theorem.

§5.2. The Total Variation

We measure the closeness of two densities f and g by their L_1 distance $\int |f-g|$. There are many reasons for this, but all more or less follow from the following observation.

THEOREM 5.1 (SCHEFFÉ'S IDENTITY). *Let f and g be two functions defined on $\mathbf{R}^d$ satisfying $\int f = \int g = 1$. Let $\mathcal{B}$ denote the class of all Borel sets of $\mathbf{R}^d$. Then*

$$\sup_{B\in\mathcal{B}} \left| \int_B f - \int_B g \right| = \frac{1}{2}\int |f-g|.$$

PROOF. Note that

$$\sup_{B\in\mathcal{B}} \left| \int_B f - \int_B g \right| = \int_{f>g} (f-g) = \int_{g>f} (g-f) = \frac{1}{2}\int |f-g|. \quad \square$$

We can thus compare the performance of density estimates on an absolute scale, and L_1 distances indeed have a physical interpretation: if we know that $\int |f-g| < 0.04$, then we know that differences in probabilities are at most 0.02. In contrast, the interpretation of the inequalities $\int (f-g)^2 \le 0.04$ and $\int f \log(f/g) \le 0.04$ in terms of probabilities is less evident.

§5.3. Invariance

A property that we will make frequent use of is the invariance of the total variation with respect to monotone transformations of the axes. In fact, let $T : \mathbf{R}^d \to S \subseteq \mathbf{R}^d$ be a bijection and a Borel-measurable mapping. Let X and Y be random variables with densities f and g, respectively. Then the total variation distance is

$$\sup_B |\mathbf{P}\{X \in B\} - \mathbf{P}\{Y \in B\}|.$$

If T is a bijection, this is equal to

$$\sup_B |\mathbf{P}\{T(X) \in T(B)\} - \mathbf{P}\{T(Y) \in T(B)\}|,$$

which is nothing but the total variation distance between $T(X)$ and $T(Y)$, provided that $\{T(B) : B \in \mathcal{B}\}$ coincides with the Borel sets of $S = T(\mathbf{R}^d)$. We leave this easy verification as an exercise.

The implications of this fact are tremendous. For example, if we know that a certain density estimate performs very well when the data belong to a compact set such as $[-1,1]$, then we might apply the monotone mapping $T : x \to x/(1+|x|)$, estimate the density of $T(X)$, and obtain an estimate of the density of X by well-known back-transformation methods. Such estimates are called transformed density estimates. We will work out one example in Chapter 14. Other favorite mappings include rotations to align data with the axes. For example, if we suspect that the data look Gaussian, then we might first estimate the covariance matrix, and transform the data so that if the covariance matrix had been correctly estimated, the data would be standard Gaussian. In that case, of course, we could just estimate each marginal density, and afterward apply a back-transformation. The error is basically that made on the standardized (transformed) data.

Here is another variation on the same theme: assume that we wish to visualize the densities f and g to get an idea of the size and the location of the error. Computer screens cannot show infinite tails, but by showing the graphs of the transformed densities (for a transformation to a compact set), we can make do, as the total variation distance remains unchanged.

§5.4. Mappings

The total variation distance decreases under all mappings. That is, if $T : \mathbf{R}^d \to S \subseteq \mathbf{R}^k$, where k is arbitrary, then the total variation distance between $T(X)$ and $T(Y)$ is smaller than that between X and Y:

THEOREM 5.2. *Consider a Borel measurable mapping $T : \mathbf{R}^d \to S \subseteq \mathbf{R}^k$, then for any random variables X and Y on the Borel sets of $\mathbf{R}^d$:*

$$\sup_{A \in \mathcal{B}_d} |\mathbf{P}\{X \in A\} - \mathbf{P}\{Y \in A\}| \geq \sup_{A \in \mathcal{B}_k} |\mathbf{P}\{T(X) \in A\} - \mathbf{P}\{T(Y) \in A\}|,$$

where $\mathcal{B}_d$ denotes the Borel sets of $\mathbf{R}^d$.

PROOF. Note that

$$\begin{aligned}\sup_{A \in \mathcal{B}_k} &|\mathbf{P}\{T(X) \in A\} - \mathbf{P}\{T(Y) \in A\}| \\ &= \sup_{A \in \mathcal{B}_k} |\mathbf{P}\{X \in T^{\text{inv}}(A)\} - \mathbf{P}\{Y \in T^{\text{inv}}(A)\}| \\ &\leq \sup_{A \in \mathcal{B}_d} |\mathbf{P}\{X \in A\} - \mathbf{P}\{Y \in A\}|\end{aligned}$$

because $\{T^{\text{inv}}(A) : A \in \mathbf{R}^k\} \subseteq \mathcal{B}_d$. □

This theorem has interesting implications. For example, let f and g be densities on $\mathbf{R}^d$ for the random variables X and Y, and let f^* and g^* denote one pair of the k-dimensional marginal densities. Let $X = (X_1, \ldots, X_d)$ have density f. Then f^* is in fact the density of $T(X_1, \ldots, X_d) = (X_1, \ldots, X_k)$, if we assume that the marginal is for the first k components. Thus,

$$\int |f^* - g^*| \leq \int |f - g|.$$

If f^+ and g^+ are the densities of $\|X\|$ and $\|Y\|$, respectively, then

$$\int |f^+ - g^+| \le \int |f - g|.$$

Maps decrease the L_1 distance.

§5.5. Convolutions

We define the convolution density $f * K$ as the density of $X + Y$, where X has density f and Y has density K on $\mathbf{R}^d$. Think of this as a perturbation Y applied to X. We recall that

$$f * K(x) = \int f(z)K(x - z)\,dz.$$

The definition above for a convolution also holds when f and/or K are absolutely integrable functions, and in that case we have

$$\int |f * K| \le \int |f| \times \int |K|,$$

a fact that is easy to check by change of integration, and will be referred to as Young's inequality. The convolution operation too lowers the total variation distance: for any densities f, g and for any integrable function K:

$$\int |f * K - g * K| \le \int |K| \int |f - g|.$$

To see this, we merely apply Young's inequality:

$$\int |f * K - g * K| = \int |(f - g) * K| \le \int |f - g| * |K| = \int |f - g| \int |K|.$$

§5.6. Normalization

In some cases, we may end up with density estimates that do not quite integrate to one or be positive. So assume f is a density and g is an estimate. Normalize g first by replacing it by $\max(g, 0)$. This lowers the L_1 error, as clearly

$$\int |f - \max(g, 0)| < \int |f - g|.$$

Next, assuming that $g \ge 0$, we normalize the integral to make it a density by considering $g / \int g$. Then we have

$$\int \left| f - \frac{g}{\int g} \right| \le \int |f - g|$$

This is left as an exercise.

§5.7. The Lebesgue Density Theorem

Note that two densities f and g are identical if $\int_B f = \int_B g$ for all Borel sets B, that is, if and only if $\int |f - g| = 0$. Thus, we may alter f on a set of zero Lebesgue measure, and still be in the same equivalence class of densities. This immediately makes criteria such as $|f(x) - g(x)|$ (pointwise error) or $\sup_x |f(x) - g(x)|$ (the supremum error) suspect, unless we replace these criteria by quantities that are defined in terms of these equivalence classes. Some help in this respect comes from the following theorem.

THEOREM 5.3 (THE LEBESGUE DENSITY THEOREM). *Let $\mathcal{Q}$ be a subclass of the Borel sets of $\mathbf{R}^d$ with the property that*

$$\sup_{Q \in \mathcal{Q}} \frac{\lambda(Q^*)}{\lambda(Q)} \leq c < \infty$$

for some constant c, where Q^* is the smallest cube centered at the origin that contains Q, and $\lambda(\cdot)$ denotes the volume of a set. (Here $a/0$ is treated as ∞ for all $a \geq 0$.) Let $\mathcal{Q}_r$ be the subclass of $\mathcal{Q}$ containing only sets Q with $\lambda(Q) \leq r$. Let f be any density on $\mathbf{R}^d$. Let $z + Q$ denote the translation of Q by z. Then for almost all x:

$$\lim_{r \downarrow 0} \sup_{Q \in \mathcal{Q}_r} \left| \frac{1}{\lambda(Q)} \int_{x+Q} f(y)\, dy - f(x) \right| = 0.$$

The points x at which this convergence takes place are called Lebesgue points for f. Classes that satisfy the condition are the classes of all cubes, or all balls of $\mathbf{R}^d$.

This theorem is typically applied to balls $B_{x,r}$ of radius r centered at x. Let us define $g(x) = \liminf_{r \downarrow 0} \int_{B_{x,r}} f / \lambda(B_{x,r})$. The Lebesgue density theorem says that g is in the equivalence class of f. And in fact, it is easy to see that all members of that equivalence class lead to the same g, and thus, g may be considered as the representative of that class. We will call it the Lebesgue representative, denoted temporarily by $L(f)$. Now, of course, we may return to the pointwise criterion, and use $|L(f)(x) - L(g)(x)|$ instead of $|f(x) - g(x)|$. And we may use $\sup_x |L(f)(x) - L(g)(x)|$ as the new definition of supremum distance.

§5.8. LeCam's Inequality

The following inequality will be useful in Chapter 15.

THEOREM 5.4. *For any densities f, g:*

$$\int \sqrt{fg} \geq \int \min(f,g) \geq \frac{1}{2}\left(\int \sqrt{fg}\right)^2 .$$

(The second inequality is called LeCam's inequality (1973).)

PROOF. The first inequality is trivial, so we only prove the second one:

$$\begin{aligned} 2 - 2\int \min(f,g) &= 2\int \left(\frac{f+g}{2} - \min(f,g)\right) \\ &= \int (\max(f,g) - \min(f,g)) \\ &= \int |f-g| \\ &= \int |(\sqrt{f} - \sqrt{g})(\sqrt{f} + \sqrt{g})|. \end{aligned}$$

Also, using the above chain of identities, and the Cauchy–Schwarz inequality,

$$\begin{aligned} \left(2 - 2\int \min(f,g)\right)^2 &\leq \int (\sqrt{f} - \sqrt{g})^2 \int (\sqrt{f} + \sqrt{g})^2 \\ &= \left(2 - 2\int \sqrt{fg}\right)\left(2 + 2\int \sqrt{fg}\right) \\ &= 4 - 4\left(\int \sqrt{fg}\right)^2 . \end{aligned}$$

The last inequality implies that $2\int \min(f,g) \geq \left(\int \min(f,g)\right)^2 + \left(\int \sqrt{fg}\right)^2 \geq \left(\int \sqrt{fg}\right)^2$. □

§5.9. Bibliographic Remarks

Differentiation of measures, and approximation of functions by convolutions, is dealt with at length by Stein (1970), de Guzmán (1975, 1981), and Wheeden and Zygmund (1977).

§5.10. Exercises

Exercise 5.1 (TOTAL VARIATION). Without appealing to Scheffé's identity, show that the total variation defines a distance on the space of all densities on $\mathbf{R}^d$.

Exercise 5.2 Show the following variant of Scheffé's identity: if f and g are arbitrary integrable functions defined on $\mathbf{R}^d$, then

$$\sup_{B\in\mathcal{B}}\left|\int_B f-\int_B g\right|=\frac{1}{2}\left(\int|f-g|+\left|\int f-\int g\right|\right).$$

Exercise 5.3 (RELATIONSHIP WITH L_2). The space is $\mathbf{R}$.

A. Show that for any $(p,q)\in(0,2]\times(0,\infty)$, there exists a pair of densities (f,g) such that $\int|f-g|=p$, yet $\int(f-g)^2=q$. (Thus, no general inequalities between the two norms exist.)

B. Show that $\int(f-g)^2$ is not scale-invariant.

C. Show that if $f, f_n, n\geq 1$, is a sequence of densities, then $\int(f-f_n)^2\to 0$ implies $\int|f-f_n|\to 0$. HINT: Split the integral over a compact set and its complement, and use the Cauchy–Schwarz inequality on one part.

D. Show that for any density f, there exists a sequence of functions f_n with $\int(f-f_n)^2\to 0$, yet $\int|f-f_n|\to\infty$.

Exercise 5.4 (NORMALIZATION). Let f be a density, and let g be a nonnegative function. Show that normalization never hurts:

$$\int\left|f-\frac{g}{\int g}\right|\leq\int|f-g|.$$

In fact, show that if $\int g<1$, and g^* is any density with $g^*\geq g$, then $\int|g^*-f|\leq\int|g-f|$, so it really does not matter much how the missing mass is distributed to make a density!

Exercise 5.5 (CONVOLUTIONS). If K, f, g are densities on the real line, and we define $f_n=f_{n-1}*K$, $g_n=g_{n-1}*K$, $f_0=f$, $g_0=g$, then $\int|f_n-g_n|\leq\int|f-g|$ and $\int|f_n-g_n|\downarrow$ from the inequality in the text. Is it always true that $\int|f_n-g_n|\to 0$?

Exercise 5.6 (THE KULLBACK–LEIBLER DISTANCE). Some applications consider the Kullback–Leibler distance between two densities, which, when the integral exists, is $K(f,g)=\int f\log(f/g)$. It is, of course, not a proper distance. This exercise looks at certain properties. We assume throughout that f is absolutely continuous with respect to g.

A. Using Jensen's inequality, show that $K(f,g)\geq 0$.

B. Show that any functional $\int f\psi(f/g)$ is invariant under measurable bijective transformations (this includes all strictly monotone transformations).

C. Prove Pinsker's inequality (Csiszár, 1967; Kullback, 1967; Kemperman, 1969): $\int|f-g|\leq\sqrt{2K(f,g)}$. HINT: Let $A=\{f\geq g\}$, $p=\int_A f$, $q=\int_A g$. First show $\int_A f\log(f/g)\geq p\log(p/q)$. Then show $K(f,g)\geq H(p,q)=p\log(p/q)+(1-p)\log((1-p)/(1-q))$. Set

$p = q + r$, and consider H as a function of r. By Taylor series bounding, show that $H \geq 2r^2$, and conclude.

D. For large values of $\int |f - g|$, the following inequality of Bretagnolle and Huber (1979) is slightly better than Pinsker's inequality: $\int |f - g| \leq 2\sqrt{1 - \exp(-K(f,g))}$. HINT: Apply Jensen's inequality to each of $\int f \log(\min(g/f), 1)$ and $\int f \log(\max(g/f), 1)$ and conclude.

E. For every $\epsilon > 0$, find a pair (f, g) with f absolutely continuous with respect to g such that $\int |f - g| < \epsilon$, yet $K(f, g) = \infty$.

F. Show that if f_n is the n-fold product density of f, and g_n is the n-fold product density of g, then $K(f_n, g_n) = nK(f, g)$.

G. If L is any density, then $K(f * L, g * L) \leq K(f, g)$, where $*$ is the convolution operation.

H. True or false: if f and g are densities on $\mathbf{R}^d$, with f absolutely continuous with respect to g, and f_s and g_s are any s-dimensional marginals, then is f_s absolutely continuous with respect to g_s, and if so, is $K(f_s, g_s) \leq K(f, g)$?

Exercise 5.7 (JEFFREY'S DISTANCE). For $p > 0$, Jeffrey's distance, routinely referred to as the Hellinger distance, is

$$H_p(f,g) = \left(\int |f^{1/p} - g^{1/p}|^p \right)^{1/p}.$$

Clearly, H_1 is the standard L_1 distance. These distances are all more or less small together and large together.

A. Show that $H_2^2(f,g) \leq \int |f - g| \leq 2H_2(f,g)$. HINT: Write $|f - g| = |\sqrt{f} - \sqrt{g}||\sqrt{f} + \sqrt{g}|$ and apply routine bounding methods.

B. Use LeCam's inequality to conclude that $2 - \int |f - g| \geq (1 - H_2^2(f,g)/2)^2$.

C. Show that if f_n is the n-fold product density of f, and g_n is the n-fold product density of g, then $2\int \min(f_n, g_n) \geq (1 - H_2^2(f,g)/2)^{2n}$.

Exercise 5.8 (THE HARDY–LITTLEWOOD MAXIMAL FUNCTION). Define the Hardy–Littlewood maximal function for a density f on $\mathbf{R}^d$ by

$$f^*(x) = \sup_{r>0} \frac{\int_{B(x,r)} f(y)\,dy}{\int_{B(x,r)} dy},$$

where $B(x,r)$ is the open cube of center x and side length $2r$ Show that $\sup_x f^*(x) \leq \sup_x L(f)(x)$, that f^* is not integrable, and that there exists a constant C depending upon d only such that $\int_{x: f^*(x) > t} dx \leq C/t$, $t > 0$.

Exercise 5.9 (EXTENSION OF THE LEBESGUE DENSITY THEOREM). Consider a bounded function K on $\mathbf{R}^d$ of compact support. Define $K_h(x) = (1/h^d)K(x/h)$. Consider the convolution $f * K_h(x) = \int f(y)K_h(x - y)\,dy$, where f is a density on $\mathbf{R}^d$. Extend the Lebesgue density theorem by showing that for all Lebesgue points of f,:

$$\lim_{h\to 0} f * K_h(x) = L(f)(x) \int K.$$

HINT: Approximate K by weighted sums of indicators of rectangles, and show the result when K is the indicator function of a rectangle.

§5.11. References

J. Bretagnolle and C. Huber, "Estimation des densités: Risque minimax," *Zeitschrift für Wahrscheinlichkeitstheorie und verwandte Gebiete*, vol. 47, pp. 119–137, 1979.

I. Csiszár, "Information-type measures of difference of probability distributions and indirect observations," *Studia Scientiarium Mathematicarum Hungarica*, vol. 2, pp. 299–318, 1967.

M. de Guzmán, "Differentiation of Integrals in R^n," Lecture Notes in Mathematics #481, Springer-Verlag, Berlin, 1975.

M. de Guzmán, *Real Variable Methods in Fourier Analysis*, North-Holland, Amsterdam, 1981.

J. H. B. Kemperman, "On the optimum rate of transmitting information," in: *Probability and Information Theory*, vol. 89, pp. 126–169, Springer Lecture Notes in Mathematics, Springer-Verlag, Berlin, 1969.

S. Kullback, "A lower bound for discrimination information in terms of variation," *IEEE Transactions on Information Theory*, vol. IT-13, pp. 126–127, 1967.

L. LeCam, "Convergence of estimates under dimensionality restrictions," *Annals of Statistics*, vol. 1, pp. 38–53, 1973.

E. M. Stein, *Singular Integrals and Differentiability Properties of Functions*, Princeton University Press, Princeton, NJ, 1970.

R. L. Wheeden and A. Zygmund, *Measure and Integral*, Marcel Dekker, New York, 1977.

Chapter 6
Choosing a Density Estimate

§6.1. Choosing Between Two Densities

Consider the following simple situation: g_n and f_n are two density estimates, and we must select the best one, that is, $\arg\min(\int |f_n - f|, \int |g_n - f|)$. More precisely, given the sample $X_1, \ldots, X_n$ distributed according to density f, we are asked to construct a density estimate ϕ_n such that

$$\int |\phi_n - f| \approx \min\left(\int |f_n - f|, \int |g_n - f|\right).$$

This simple problem turns out to be surprisingly difficult, even if the estimates f_n and g_n are fixed densities, not depending on the data.

The solution we propose here is fundamental, and is at the heart of many estimates discussed subsequently. First we introduce a few names and terms to save space later. The empirical measure $\mu_n(A)$ of $\mu(A) \stackrel{\text{def}}{=} \int_A f$ is the measure that gives mass $\mu_n(A) = (1/n)\sum_{i=1}^n \mathbf{1}_{[X_i \in A]}$ to set A. The set $A = A(f_n, g_n) = \{x : f_n(x) > g_n(x)\}$ will be called the *Scheffé set* for the ordered pair (f_n, g_n), as we recall Scheffé's identity,

$$\int |f_n - g_n| = 2\int (f_n - g_n)_+ = 2\int_A f_n - 2\int_A g_n,$$

valid whenever $\int f_n = \int g_n = 1$.

We define the *Scheffé estimate* f_n^* as follows:

$$f_n^* = \begin{cases} f_n & \text{if } |\int_A f_n - \mu_n(A)| < |\int_A g_n - \mu_n(A)|, \\ g_n & \text{otherwise.} \end{cases}$$

Note that this is not a symmetric definition in f_n and g_n! The situations in which we wish to carry out a selection are innumerable: we may want to pick the best of two bandwidths in kernel estimates, or the best of two summation sizes in series or wavelet estimates, or the best of a histogram and a kernel estimate, or the best of a parametric gamma density estimate and a nonparametric unimodal Grenander estimate.

THEOREM 6.1. *Let f_n and g_n be two density estimates with $\int f_n = \int g_n = 1$. For the Scheffé estimate f_n^*, we have*

$$\int |f_n^* - f| \le 3 \min\left(\int |f_n - f|, \int |g_n - f|\right) + 4 \max_{A \in \mathcal{A}} \left| \int_A f - \mu_n(A) \right|,$$

where $\mathcal{A} = \{\{f_n > g_n\}, \{g_n > f_n\}\}$.

PROOF. Denote by ξ_n the best density estimate:

$$\xi_n = \begin{cases} f_n & \text{if } \int |f_n - f| < \int |g_n - f|, \\ g_n & \text{otherwise.} \end{cases}$$

Clearly,

$$\begin{aligned} \int |f_n^* - f| &\le \int |\xi_n - f| + \int |\xi_n - f_n^*| \\ &= \int |\xi_n - f| + \int |f_n - g_n| \mathbf{1}_{[\xi_n \equiv f_n, f_n^* \equiv g_n]} \\ &\qquad + \int |f_n - g_n| \mathbf{1}_{[\xi_n \equiv g_n, f_n^* \equiv f_n]} \\ &\stackrel{\text{def}}{=} I + II + III. \end{aligned}$$

We consider II. Let $E = [\xi_n \equiv f_n, f_n^* \equiv g_n]$ and $A = A(f_n, g_n)$:

$$\begin{aligned} II &= \int |f_n - g_n| \mathbf{1}_E \\ &= 2\left(\int_A f_n - \int_A g_n\right) \mathbf{1}_E \\ &= 2\left(\int_A f_n - \mu_n(A)\right) \mathbf{1}_E - 2\left(\int_A g_n - \mu_n(A)\right) \mathbf{1}_E \\ &\le 4\left(\int_A f_n - \mu_n(A)\right) \mathbf{1}_E \\ &\le 4 \int_A (\xi_n - f) \mathbf{1}_{[E]} + 4\left(\int_A f - \mu_n(A)\right) \mathbf{1}_{[E]} \\ &\le 2 \int |\xi_n - f| + 4 \left| \int_A f - \mu_n(A) \right|. \end{aligned}$$

For the middle step above, consider the cases $\int_A f_n > \mu_n(A) > \int_A g_n$ and $\int_A f_n > \int_A g_n > \mu_n(A)$ separately. Similarly, by switching f_n and g_n, we see that

$$III \le 2 \int |\xi_n - f| + 4 \left| \int_{A(g_n, f_n)} f - \mu_n(A(g_n, f_n)) \right|.$$

Thus,

$$\int |f_n^* - f| \le 3 \min\left(\int |f_n - f|, \int |g_n - f| \right) + 4 \sup_{A \in \mathcal{A}} \left| \int_A f - \mu_n(A) \right|. \quad \square$$

The Scheffé estimate thus has an error that is within

$$E_n \stackrel{\text{def}}{=} 4 \sup_{A \in \mathcal{A}} \left| \int_A f - \mu_n(A) \right|$$

of three times the best possible error. Note that this exceedance E_n only depends upon f_n and g_n through the shape of the set A.

§6.2. Examples

Assume that f_n and g_n are both normal densities (perhaps determined by the data), and we must decide between them. Then A is either of the form (a, b) or $[a, b]^c$, and thus, $\mathcal{A}$ is contained in the collection $\mathcal{B}_2$ of unions of two intervals. But then,

$$\sup_{A \in \mathcal{B}_2} \left| \int_A f - \mu_n(A) \right| \le 2 \sup_x |F_n(x) - F(x)|,$$

where F is the distribution function for F, and F_n is the empirical distribution function. But we know from Massart's tightening of the Dvoretzky–Kiefer–Wolfowitz inequality (Section 3.3) that

$$\mathbf{E}\left\{ \sup_x |F_n(x) - F(x)| \right\} \le \frac{1}{\sqrt{n}}.$$

Therefore, if we use the empirical measure μ_n, using *the same data* that were used in obtaining f_n and g_n, regardless of the dependence of f_n and g_n upon these data, we have

$$\mathbf{E}\left\{ \int |f_n^* - f| - 3 \min\left(\int |f_n - f|, \int |g_n - f| \right) \right\} \le \frac{8}{\sqrt{n}}$$

for all n. While the coefficient "3" may not be acceptable to some, we should stress nevertheless the simplicity of the procedure. Indeed, the set $A = \{f_n > g_n\}$ is very easily determined by the solution of a quadratic equation, and $\mu_n(A)$ is indeed easy to compute. Furthermore, the inequality above applies to any normal density estimate, with arbitrary dependence on the data, and to all densities.

Consider a more sophisticated situation, in which a normal density estimate f_n is challenged by a histogram estimate g_n having k bins (of data-dependent widths, and data dependent heights in the bins). Then both

$A = A(f_n, g_n)$ and $A(g_n, f_n)$ consist of unions of at most $k+2$ disjoint intervals (we call the collection of such sets $\mathcal{B}_{k+2}$). Thus, by the argument presented above,

$$\sup_{A \in \mathcal{B}_{k+2}} \left| \int_A f - \mu_n(A) \right| \leq (k+2) \sup_x |F_n(x) - F(x)|.$$

Therefore, by Massart's inequality (see Section 3.3),

$$\begin{aligned} &\mathbf{E}\{\int |f_n^* - f| - 3 \min \left(\int |f_n - f|, \int |g_n - f| \right)\} \\ &\quad \leq 4(k+2)\mathbf{E}\left\{ \sup_x |F_n(x) - F(x)| \right\} \\ &\quad \leq \frac{4(k+2)}{\sqrt{n}}. \end{aligned}$$

Alternatively, we could have used a combination of Theorems 3.2 and 4.3 to obtain an upper bound of the form $c\sqrt{k/n}$ for some constant c. In any case, this inequality, while useful, is much too weak in general, when k grows with n. This may be explained by noting its generality: it remains valid if a mean-spirited adversary is allowed to select the bins and bin heights in the worst possible manner so as to produce the worst possible selection f_n^*. If $\mu_n(A)$ in the definition of f_n^* was based upon an independently drawn sample of size m, then in the two examples above, $\mathcal{A}$ would consist of two sets, and by the Cauchy–Schwarz inequality, for any fixed $A \in \mathcal{A}$:

$$\begin{aligned} \mathbf{E}\left\{ \left| \mu_n(A) - \int_A f \right| \,\middle|\, X_1, \ldots, X_n \right\} &\leq \sqrt{\mathbf{E}\left\{ \left(\mu_n(A) - \int_A f \right)^2 \,\middle|\, X_1, \ldots, X_n \right\}} \\ &= \mathbf{V}\left\{ \mu_n(A) | X_1, \ldots, X_n \right\} \leq \sqrt{\frac{1}{4m}}. \end{aligned}$$

Conditioning and then unconditioning, we have

$$\mathbf{E}\left\{ \int |f_n^* - f| - 3 \min \left(\int |f_n - f|, \int |g_n - f| \right) \right\} \leq 4\sqrt{\frac{1}{m}}.$$

We must think of m as a part of the data set aside for testing purposes, and will call these data test data. Typically, of course, m is much smaller than n. In the former example (normal versus normal), this test data approach does not help. However, in the second example (normal versus any-histogram), the obtained bound is better than the previous one as soon as $m/n \geq 1/(k+2)^2$. So we distinguish between μ_n based on the original data and μ_n based on test data. Mostly, for simple parametric density estimates, the original data approach is best.

§6.3. Is the Factor of Three Necessary?

The Scheffé estimate guarantees an L_1 error not more than three times the smallest error plus the absolute deviation of the empirical measure over the set $\{f_n > g_n\}$. The factor of three is clearly unacceptable if the best of the two estimates f_n and g_n has a large error. It would be desirable to get rid of any constant factor larger than one. Unfortunately, this is impossible for the Scheffé estimate.

FACT 6.1. *For any $t < 1$, $\epsilon > 0$, there exist three densities, g, h, f, such that if $f_n = h$, $g_n = g$, then the Scheffé estimate f_n^* based upon a sample of size n drawn from f has the following lower bound:*

$$\frac{\lim_{n\to\infty} \mathbf{E}\{\int |f_n^* - f|\}}{\min\left(\int |g-f|, \int |h-f|\right)} > 2 - \epsilon.$$

Furthermore, $\lim_{n\to\infty} \mathbf{E}\left\{\int |f_n^* - f|\right\} = t$.

PROOF. Just consider this simple example. Call our collection of density estimates $\mathcal{F} = \{h, g\}$. Take five disjoint sets A, B, C, D, E on $\mathbf{R}^d$ of volumes $1-b-c, b, c, b+c, c$ each, where $0 < b < c$, $b+c < 1$. Let h be uniform on $A \cup B \cup C$. Let g be uniform on $A \cup D$. Let f be uniform on $A \cup B \cup E$. Clearly, $\int |f-h| = 2c < 2(b+c) = \int |f-g|$, so h is closer to the true density f. However, Scheffé's estimate picks g for all n large enough: indeed, the Scheffé set $\{h > g\}$ is $B \cup C$. We verify that the mass of this set under h, g and f is $b+c$, 0, and b, respectively, Clearly, if the sample size n is infinite, then $\mu_n(H)$ may be replaced by $\int_H f$ in the definition of the Scheffé density estimate. As $b < c$, we have $f_n^* = g$ for all n large enough by the strong law of large numbers. Thus,

$$\lim_{n\to\infty} \mathbf{E}\left\{\int |f_n^* - f|\right\} = \int |g-f| = 2(b+c).$$

Yet,

$$\min\left(\int |g-f|, \int |h-f|\right) = \int |h-f| = 2c.$$

The ratio of these two quantities is $(b+c)/c$, which can be made as close to 2 as desired by the choice of c and $b < c$. □

In spite of the necessity of a constant factor larger than one, the Scheffé estimate is the only one we know for which the estimate is guaranteed to be within a constant factor of the smaller error plus a uniformly conrollable term. Below we point out that other popular estimates do not possess such a universal property

§6.4. Maximum Likelihood Does not Work

One may wonder why we decided not to use another method of selection, such as the maximum likelihood. This is simply because the maximum likelihood criterion, even when applied under idealized circumstances, does not minimize L_1. Consider the real line, with two density estimates: $f_0 = \frac{1}{2}\mathbf{1}_{[-1,1]}$ and $g_\delta = \mathbf{1}_{[\delta,1+\delta]}$, where $\delta \in (0,1)$ is to be picked. Assume that the true density f is the uniform density on $[0,1]$. If the data are $X_1,\ldots,X_n$, the maximum likelihood choice ϕ_n would be defined by

$$\phi_n \stackrel{\text{def}}{=} \arg\max\left(\prod_{i=1}^n f_0(X_i), \prod_{i=1}^n g_\delta(X_i)\right).$$

Clearly, everything depends upon N, the number of data points in $[0,\delta)$, a binomial (n,δ) random variable. We note that g_δ is picked if and only if $N=0$. Also, $\int|f_0-f| = 1$, $\int|g_\delta - f| = 2\delta$. Let $\delta < \frac{1}{2}$ so that g_δ is the better estimate. Clearly,

$$\begin{aligned}\mathbf{E}\left\{\int|\phi_n - f|\right\} - \int|g_\delta - f| &= (1-2\delta)\mathbf{P}\{N>0\} \\ &= (1-2\delta)(1-(1-\delta)^n) \\ &\to 1-2\delta,\end{aligned}$$

as $n\to\infty$. Thus, while for small δ the expected error for the best density is a respectable 2δ, the maximum likelihood estimate makes, with probability tending to one, a catastrophic blunder.

§6.5. $\mathbf{L_2}$ Distances Are To Be Avoided

Others may want to minimize $\int(f_n - f)^2$, the L_2 distance. But even if we were given f, and even if that f were square integrable, we might still make catastrophic choices. Consider this situation: f is once again the uniform density on $[0,1]$, and we are indeed given this information. Let ϵ be a small positive number, and assume that both density estimates ignore the data: f_ϵ is our catastrophic candidate—it is $1-\epsilon$ on $[0,1]$, and ϵ^3 on $[1,1+1/\epsilon^2]$ (a long skinny tail, if you wish). Now, $g_\epsilon = 1$ on $[-\epsilon^2, 1-\epsilon^2]$. Verify that $\int|g_\epsilon - f| = 2\epsilon^2 < 2\epsilon = \int|f_\epsilon - f|$, so that the choices are not even close! However, $\int(f_\epsilon - f)^2 = \epsilon^2+\epsilon^4 < 2\epsilon^2 = \int(g_\epsilon - f)^2$, so that minimizing the L_2 error, even with f given, picks the wrong density from the set $\{f_\epsilon, g_\epsilon\}$! Therefore, for universal properties in density estimation, criteria that are based on the square integrated distance are doomed. In contrast, the Scheffé estimate does not make such bad mistakes.

§6.6. Selection from k Densities

Consider now the selection problem with k candidates $f_{ni}, 1 \le i \le k$, such that $\int f_{ni} = 1$ for all i. Let $A_{ij}, i < j$, denote the Scheffé set $A(f_{ni}, f_{nj}) = \{x : f_{ni} > f_{nj}\}$. We say that f_{ni} wins against f_{nj} when

$$\left| \int_{A_{ij}} f_{ni} - \mu_n(A_{ij}) \right| < \left| \int_{A_{ij}} f_{nj} - \mu_n(A_{ij}) \right| .$$

Of course, for each pair $i < j$, if f_n^* denotes the winner, we have by Theorem 1:

$$\begin{aligned}\int |f_n^* - f| & \\ &\le 3 \min \left(\int |f_{ni} - f|, \int |f_{nj} - f| \right) + 4 \sup_{A \in \{A_{ij}, A_{ji}\}} \left| \int_A f - \mu_n(A) \right| \\ &= 3 \min \left(\int |f_{ni} - f|, \int |f_{nj} - f| \right) + 4\Delta,\end{aligned}$$

where

$$\Delta = \sup_{A \in \{A_{ij}, A_{ji}; 1 \le i < j \le k\}} \left| \int_A f - \mu_n(A) \right| .$$

Let us thus run a little competition with $k(k-1)/2$ matches, one for each ordered pair $i < j$. For each f_{ni}, we total the number of wins, and declare the density estimate with the maximum number of wins the Scheffé tournament winner. Ties are broken by picking the smallest index.

THEOREM 6.2. *For the Scheffé tournament winner f_n^*, we have*

$$\int |f_n^* - f| \le 9 \min_i \int |f_{ni} - f| + 16\Delta.$$

PROOF. Let $m = \min_i \int |f_{ni} - f|$. Group the density estimates as follows: in group 0, we place all those with $\int |f_{ni} - f| = m$; in group 1, those with $\int |f_{ni} - f| \in (m, 3m + 4\Delta]$; in group 2, those with $\int |f_{ni} - f| \in (3m + 4\Delta, 9m + 16\Delta]$; in group 3, those with $\int |f_{ni} - f| > 9m + 16\Delta$. Let the number of density estimates in each group be denoted by n_0, n_1, n_2, and n_3, respectively. From Theorem 1, it is clear that any group 0 estimate must win against any group 2 or 3 estimate, and thus, their number of wins is at least $n_2 + n_3$. Any group 3 estimate must lose to any group 0 or 1 estimate, and thus, its number of wins is at most $n_2 + n_3 - 1$. Therefore, the Scheffé tournament winner cannot be from group 3. □

We may also declare another winner. We define

$$\Delta_i = \sup_{A \in \mathcal{A}} \left| \int_A f_{ni} - \mu_n(A) \right| ,$$

where $\mathcal{A} = \{A_{ij}, A_{ji} : 1 \le i < j \le k\}$. The minimum distance estimate ψ_n is that f_{ni} of smallest index that minimizes Δ_i. It is called a minimum distance estimate because it minimizes the distance to the empirical measure in a metric that is reminiscent of the total variation or L_1 distance—it would be the total variation distance if $\mathcal{A}$ were replaced by the class of Borel sets, $\mathcal{B}$. Only, the total variation distance

$$\sup_{B \in \mathcal{B}} \left| \int_B f - \mu_n(B) \right| = 1$$

for any density f and any empirical measure μ_n: just let $B = \{X_1, \ldots, X_n\}$, for example. If each computation of $\int_A f_{ni}$ requires one time unit, then the computation of the Scheffé tournament winner requires $k(k-1)/2$ time units, while that for the minimum distance estimate requires $k^2(k-1)/2$ time units—it is about k times computationally more intensive. This extra cost of course draws benefits, as is shown in the next theorem.

THEOREM 6.3. *For the minimum distance estimate ψ_n, we have*

$$\int |\psi_n - f| \le 3 \min_i \int |f_{ni} - f| + 4\Delta.$$

PROOF. Let $\psi_n = f_{ni}$ and let f_{nj} be any density estimate minimizing $\int |f_{n\ell} - f|$ over all ℓ. Assume that $j \ne i$. Then, clearly,

$$\int |\psi_n - f| \le \int |f_{nj} - f| + \int |f_{ni} - f_{nj}|.$$

Now, assuming without loss of generality that $i < j$:

$$\begin{aligned}
\int |f_{ni} - f_{nj}| &= 2 \sup_{A \in \{A_{ij}, A_{ji}\}} \left| \int_A f_{ni} - \int_A f_{nj} \right| \\
&\le 2 \sup_{A \in \mathcal{A}} \left| \int_A f_{ni} - \int_A f_{nj} \right| \\
&\le 2 \sup_{A \in \mathcal{A}} \left| \int_A f_{ni} - \mu_n(A) \right| + 2 \sup_{A \in \mathcal{A}} \left| \int_A f_{nj} - \mu_n(A) \right| \\
&\le 4 \sup_{A \in \mathcal{A}} \left| \int_A f_{nj} - \mu_n(A) \right| \quad \text{(by definition of } \psi_n = f_{ni}) \\
&\le 4 \sup_{A \in \mathcal{A}} \left| \int_A f_{nj} - \int_A f \right| + 4 \sup_{A \in \mathcal{A}} \left| \int_A f - \mu_n(A) \right| \\
&\le 4 \sup_{B \in \mathcal{B}} \left| \int_B f_{nj} - \int_B f \right| + 4\Delta \\
&= 2 \int |f_{nj} - f| + 4\Delta
\end{aligned}$$

which together with the first inequality is all that is needed. □

§6.7. Examples Continued

If we continue our examples, we note that if we are given any number k of normal density estimates, then the class $\mathcal{A} \subseteq \mathcal{B}_2$, the class of unions of two intervals. Thus, we have

$$\sup_{A \in \mathcal{B}_2} \left| \int_A f - \mu_n(A) \right| \leq 2 \sup_x |F_n(x) - F(x)|,$$

where F is the distribution function for F, and F_n is the empirical distribution function, we have, for $t > 0$, if μ_n is the standard empirical measure,

$$\begin{aligned} \mathbf{E}\left\{ \sup_{A \in \mathcal{A}} \left| \int_A f - \mu_n(A) \right| \right\} &\leq \mathbf{E}\left\{ \sup_{A \in \mathcal{B}_2} \left| \int_A f - \mu_n(A) \right| \right\} \\ &\leq \mathbf{E}\left\{ 2 \sup_x |F_n(x) - F(x)| \right\} \\ &\leq \frac{2}{\sqrt{n}}. \end{aligned}$$

Thus, for the minimum distance estimate ψ_n:

$$\mathbf{E}\left\{ \int |\psi_n - f| - 3 \min_i \int |f_{ni} - f| \right\} \leq \frac{8}{\sqrt{n}}.$$

Note, in particular, that this bound does not depend upon k. In fact, we may select the best among k arbitrary normal density estimates, and regardless of the value of k, we will obtain the same performance guarantees. In fact, Theorem 3 applies also to the infinite selection problem, although computations become a nightmare. With a selection from a possibly uncountably infinite class of densities, we can no longer index by integers i, and we can no longer be sure that minima are attained. The set-up must therefore be slightly modified. Also, we can no longer run Scheffé tournaments. Luckily, the minimum distance method remains valid.

§6.8. Selection from an Infinite Class

Let our density estimates be parametrized by $\theta \in \Theta$, where θ could represent the mean–variance pair in a normal density estimate, or the bandwidth in a kernel estimate, or the number of terms in a wavelet estimate. Let $f_{n,\theta}$ denote the density estimate with parameter θ. Let

$$\mathcal{A} = \{\{f_{n,\theta} > f_{n,\theta'}\} : \theta \in \Theta, \theta' \in \Theta, \theta \neq \theta'\}.$$

Define

$$\Delta_\theta = \sup_{A \in \mathcal{A}} \left| \int_A f_{n,\theta} - \mu_n(A) \right|.$$

Let ψ_n be a density estimate selected from among those density estimates $f_{n,\theta}$ with $\Delta_\theta < \inf_{\theta^* \in \Theta} \Delta_{\theta^*} + 1/n$. The $1/n$ here is added to ensure the existence of such a density estimate.

THEOREM 6.4. *Let $\{f_{n,\theta} : \theta \in \Theta\}$ be an arbitrary class of density estimates satisfying $\int f_{n,\theta} = 1$. For the minimum distance estimate ψ_n as defined above for an infinite selection problem, we have*

$$\int |\psi_n - f| \leq 3 \inf_{\theta \in \Theta} \int |f_{n,\theta} - f| + 4\Delta + \frac{3}{n},$$

where $\Delta = \sup_{A \in \mathcal{A}} \left| \int_A f - \mu_n(A) \right|$.

PROOF. Mimic the proof of Theorem 3. The details are left to the reader as an easy exercise. □

A variant of Theorem 4 that does not require the candidate estimates to integrate to one is given as an exercise. All theorems above make a crucial connection between the error of the selected density estimate and the supremum error Δ for empirical probability measures over certain classes of sets. This permits us thus to use the rich theory of the uniform convergence of empirical measures initiated by Vapnik and Chervonenkis to density estimation.

§6.9. Bibliographic Remarks

Scheffé's lemma (Scheffé, 1947) was at the basis of our approach. The idea for selecting sets of the form $f > g$ when dealing with L_1 norms in density estimation was grabbed from Yatracos (1985) and developed by Devroye and Lugosi (1996, 1997) in the context of kernel density estimation.

§6.10. Exercises

Exercise 6.1 Prove Theorem 4.

Exercise 6.2 Apply Exercise 5.2 to prove the following variant of Theorem 4. Let $\{f_{n,\theta} : \theta \in \Theta\}$ be an arbitrary class of density estimates (not necessarily satisfying $\int f_{n,\theta} = 1$). Let ψ_n be the minimum distance estimate defined in Section 8. Prove that

$$\int |\psi_n - f| \leq 5 \inf_{\theta \in \Theta} \int |f_{n,\theta} - f| + 4\Delta + \frac{5}{n},$$

§6.11. References

L. Devroye and G. Lugosi, "A universally acceptable smoothing factor for kernel density estimation," *Annals of Statistics*, vol. 24, pp. 2499–2512, 1996.

L. Devroye and G. Lugosi, "Nonasymptotic universal smoothing factors, kernel complexity and Yatracos classes," *Annals of Statistics*, vol. 25, pp. 2626–2637, 1997.

A. Dvoretzky, J. Kiefer, and J. Wolfowitz, "Asymptotic minimax character of a sample distribution function and of the classical multinomial estimator," *Annals of Mathematical Statistics*, vol. 33, pp. 642–669, 1956.

P. Massart, "The tight constant in the Dvoretzky–Kiefer–Wolfowitz inequality," *Annals of Probability*, vol. 18, pp. 1269–1283, 1990.

H. Scheffé, "A useful convergence theorem for probability distributions," *Annals of Mathematical Statistics*, vol. 18, pp. 434–458, 1947.

Y. G. Yatracos, "Rates of convergence of minimum distance estimators and Kolmogorov's entropy," *Annals of Statistics*, vol. 13, pp. 768–774, 1985.

Chapter 7
Skeleton Estimates

§7.1. Kolmogorov Entropy

In this short chapter, we show how to pick a density estimate from a prespecified class of densities, $\mathcal{F}$. In particular, the classes we are interested in are totally bounded, that is, for every $\epsilon > 0$, there exists a finite number N_ϵ of densities in $\mathcal{F}$ such that the L_1 balls of radius ϵ centered at these densities cover $\mathcal{F}$, that is, if these chosen densities are $\mathcal{G}_\epsilon = \{g_1, \ldots, g_{N_\epsilon}\}$, then

$$\mathcal{F} \subseteq \bigcup_{i=1}^{N_\epsilon} B_{g_i,\epsilon},$$

where $B_{g,r} = \{f : \int |f - g| \leq r\}$. The smallest such N_ϵ is called the *complexity* (or *covering number*) of $\mathcal{F}$ (and thus is a function of ϵ), and $\log_2 N_\epsilon$ is the *Kolmogorov entropy* of $\mathcal{F}$. $\mathcal{G}_\epsilon$ is called a *skeleton* of $\mathcal{F}$.

§7.2. Skeleton Estimates

We define a skeleton density estimate f_n as any density estimate with the property that $f_n \in \mathcal{G}_\epsilon$ for some ϵ. The size of $\mathcal{G}_\epsilon$ is typically unwieldy, and so the original interest in such estimates was purely theoretical. However, as $\mathcal{G}_\epsilon$ is a finite set, and usually one with a lot of structure, the information-theoretic aspect of a skeleton should not be lost on the reader. A density estimate can now be identified by an integer, and in a sense such estimates are prime candidates for compressing data! With a good skeleton density estimate, we may throw away all the data and just store an integer between 1 and N_ϵ. This would require about $\log_2 N_\epsilon$ bits, and this explains of course why we call this quantity the Kolmogorov entropy. In fact, because of this important connection with data compression, we feel that much more research is needed in this area of density estimation.

Now, on to the task at hand—how will we pick $f_n \in \mathcal{G}_\epsilon$? We could achieve this by our minimum distance principle. To this end, for fixed $\epsilon > 0$, define the Yatracos class

$$\mathcal{A}_\epsilon = \{\{x : f(x) > g(x)\}; f, g \in \mathcal{G}_\epsilon\}.$$

This class has cardinality not exceeding N_ϵ^2. Consider now the minimum distance estimate proposed by Yatracos (1985):

$$f_n = \arg\min_{f \in \mathcal{G}_\epsilon} \sup_{A \in \mathcal{A}} \left| \int_A f - \mu_n(A) \right| ,$$

where $\mu_n(A) = (1/n) \sum_{i=1}^n 1_{[X_i \in A]}$ is the empirical measure of A based on the data $X_1, \ldots, X_n$. We call f_n the *skeleton estimate.*

THEOREM 7.1. *Assume that $f \in \mathcal{F}$. For the skeleton estimate f_n determined by the minimum distance method,*

$$\mathbf{E}\left\{ \int |f_n - f| \right\} \leq 3\epsilon + \sqrt{\frac{8 \log (2N_\epsilon^2)}{n}},$$

where $\epsilon > 0$ is such that the Kolmogorov entropy N_ϵ is at least two.

PROOF. By Theorem 6.3, we have for any density f:

$$\int |f_n - f| \leq 3 \min_{g \in \mathcal{G}_\epsilon} \int |g - f| + 4 \sup_{A \in \mathcal{A}} \left| \int_A f - \mu_n(A) \right| .$$

If $f \in \mathcal{F}$, we bound this further as follows:

$$\int |f_n - f| \leq 3\epsilon + 4 \sup_{A \in \mathcal{A}} \left| \int_A f - \mu_n(A) \right| .$$

A direct application of Lemmas 2.1 and 2.2 shows that

$$\mathbf{E}\left\{ \sup_{A \in \mathcal{A}} \left| \int_A f - \mu_n(A) \right| \right\} \leq \sqrt{\frac{\log(2N_\epsilon^2)}{2n}},$$

concluding the proof. □

Observe that regardless of how fast N_ϵ grows as $\epsilon \to 0$, we may always choose an $\epsilon = \epsilon_n$ such that the expected L_1 error of the estimate converges to zero *uniformly,* as long as $\mathcal{F}$ is totally bounded. For example, $\epsilon_n = \inf\{\epsilon : \log N_\epsilon \leq \sqrt{n}\}$ will do. The uniform rate of convergence over $\mathcal{F}$ is fundamentally influenced by the Kolmogorov entropy. It is now a matter of minimizing the bound with respect to ϵ to obtain actual rates of convergence. Note that up to this point, all bounds are nonasymptotic in nature. The remainder of this chapter is used to work out some bounds on the Kolmogorov entropy, and to assess its impact on density estimation for certain classes.

§7.3. Robustness

In Theorem 1 we assume that the true density is in the class $\mathcal{F}$. Such assumptions are often made based on a statistical model. Unfortunately, the validity of such a model is very difficult to verify. Perhaps the main strength of the minimum distance method is its robustness. Indeed, without any assumption of the true density f, for the skeleton estimate we have

$$\begin{aligned}
\mathbf{E}\int |f_n - f| &\leq 3 \min_{g\in\mathcal{G}_\epsilon} \int |g-f| + 4\mathbf{E}\left\{\sup_{A\in\mathcal{A}}\left|\int_A f - \mu_n(A)\right|\right\} \\
&\qquad \text{(by Theorem 6.3)} \\
&\leq 3\min_{g\in\mathcal{F}} \int |g-f| + 3\epsilon + 4\mathbf{E}\left\{\sup_{A\in\mathcal{A}}\left|\int_A f - \mu_n(A)\right|\right\}, \\
&\qquad \text{(by the definition of } \mathcal{G}_\epsilon \text{, and the triangle inequality),} \\
&\leq 3\min_{g\in\mathcal{F}} \int |g-f| + 3\epsilon + \sqrt{\frac{8\log(2N_\epsilon^2)}{n}}.
\end{aligned}$$

The above inequality shows that modeling errors are penalized by an error proportional to the distance of the true density to the class $\mathcal{F}$. If f lies in the vicinity of $\mathcal{F}$, the obtained bound is still acceptable.

§7.4. Finite Mixtures

Let us take the simple class $\mathcal{F}$ consisting of all densities that are convex combinations of k fixed densities $g_1, \ldots, g_k$, that is, all densities of the form $\sum_{i=1}^k p_i g_i$, where the p_i's form a probability vector. Clearly, a skeleton set $\mathcal{G}_\epsilon$ is trivially constructed. Let us take all convex combinations in which each p_i is a multiple of $1/\ell$ where ℓ is a fixed integer. Then, if f is defined by $(p_1, \ldots, p_k)$, find an element in the skeleton by first finding the largest integers r_i such that $r_i/\ell \leq p_i$. Then increase r_1, r_2, and so forth, in turn by one until for the first time $\sum_i r_i/\ell = 1$. The skeleton density for $(r_1/\ell, \ldots, r_k/\ell)$ is such that $|p_i - r_i/\ell| \leq 1/\ell$. Thus, the L_1 distance between f and this density is at most k/ℓ, by the triangle inequality. As this must be smaller than ϵ, we set $\ell = \lceil k/\epsilon \rceil$, and note that N_ϵ is the number of integer vectors $(r_1, \ldots, r_k)$ with nonnegative values and total sum ℓ. We count this easily as follows: start with ℓ matchsticks on a table, which will be separated into k groups by $k-1$ pencils placed between them. In total, there are $\ell + k - 1$ objects on the table. The number of different partitions clearly is the number of ways of choosing the $k-1$ positions for the pencils, or $\binom{\ell+k-1}{k-1}$. Thus,

$$N_\epsilon \leq \binom{\lceil k/\epsilon\rceil + k - 1}{k-1} \leq \left(k + \frac{k}{\epsilon}\right)^k.$$

The Kolmogorov entropy is $O(k \log(1/\epsilon))$ as $\epsilon \downarrow 0$. This rate for the Kolmogorov entropy, polynomial in $\log(1/\epsilon)$, is typical of small parametric classes. It is easily seen from the bound of Theorem 1 that the best choice for ϵ is about $\sqrt{k/n}$. Resubstitution in the bound of Theorem 1 then shows that

$$\sup_{f \in \mathcal{F}} \mathbf{E}\left\{\int |f_n - f|\right\} \leq \sqrt{\frac{Ck \log n}{n}}$$

for a universal constant C not depending upon k, n or the densities g_i. The $\log n$ factor can be avoided, but not by the methodology given here (see Section 8.2). More importantly, the explicit bounds are useful for situations in which parameters such as k grow with n. What is the uniform rate when $k = \sqrt{n}$, for example? One may simply plug this value in the above bound to conclude a uniform error rate $O(\sqrt{\log n}/n^{1/4})$. It is in such situations that explicit bounding methods shine.

§7.5. Monotone Densities on the Hypercube

Let us next take a class that is slightly more complex, and perhaps what most would call nonparametric. Let $\mathcal{F}$ be the class of all densities on $[0,1]^d$, monotonically decreasing, and bounded by B, which we assume to be an integer at least equal to 1. That is, each density is decreasing in any coordinate if all other coordinates are held fixed. These classes are totally bounded, and we may indeed construct simple skeletons. The next lemma provides an upper bound for the Kolmogorov entropy of $\mathcal{F}$. Its implications are discussed at the end of this section.

LEMMA 7.1. *For $\mathcal{F}$, and all $\epsilon \leq 1$:*

$$\log_2 N_\epsilon \leq \frac{CB^d}{\epsilon^d},$$

where the constant C depends only on d.

PROOF. Partition the axes (that is, $[0,1]$ for each coordinate) into k equal intervals, and partition the range for the densities ($[0,B]$) into ℓ equal intervals of length B/ℓ each, where k and ℓ are integers. Let our candidate skeleton $\mathcal{G}_\epsilon$ consist of all monotone densities that are constant on each product of intervals (so, there are k^d such grid cells on which these densities are constant), and whose values on these intervals are integer multiples of B/ℓ. If $r_i B/\ell$ is the value on the ith cell in the grid with r_i integer, then the requirement that we have densities means that $B\sum_i r_i = \ell k^d$. We must do two things: (1) Compute and upper bound for $|\mathcal{G}_\epsilon|$; (2) Prove that $\mathcal{G}_\epsilon$ is indeed a skeleton for $\mathcal{F}$.

COMPUTATION OF THE CARDINALITY OF $\mathcal{G}_\epsilon$. A quick-triggered reader may be tempted to bound the complexity of this class by $(\ell+1)^{k^d}$ as for each cell, we assign one of $\ell+1$ possible values. However, that is not good enough as we will soon discover. Instead, we count much more carefully. Consider the $k^d \times \ell$ cells of $\mathbf{R}^{d+1}$ formed by the range and the domain. Index these cells by $d+1$ integers in the obvious manner. For a fixed density in the skeleton, all these cells are colored white, except those that entirely fit under the curve of the skeleton density, which are painted black. By monotonicity, if a cell is black, then so must be all cells all of whose indices are smaller. Thus, this density is totally described if for any collection of d indices out of the $d+1$, and for any value for that collection, we know the maximal index of the missing coordinate of any black cell with the given d coordinates. That is, if $(i_1, \ldots, i_m, \ldots, i_{d+1})$ is fixed except for i_m, then we need to know the maximal value i_m can take for any black cell. Mark all these extremal cells. Mark then all neighbors of these cells (note that each cell has $2d+2$ neighbors at most). The marked cells form a connected set: each marked cell is the neighbor of at least one other marked set. We visualize this set as a graph in which an edge connects neighboring cells. This graph has at most $(2d+2)(k^d + d\ell k^{d-1})$ nodes. The number of possible collections of marked cells is counted by counting the number of possible depth first search traversals of this graph started at the highest marked cell near the origin (for depth first search, consult, e.g., Tarjan, 1983). In this traversal, neighboring nodes are always visited in lexicographical ordering. Describe the depth first search path by the last coordinate of the starting point (ℓ choices here) and the sequence of indices of the neighbors followed, where a neighbor index is an integer between 1 and $2d+2$ (as in turtle graphics, where we are instructed to go north, then east, then south, then north again, and so forth). The length of a depth first search path is not more than two times the number of nodes in a graph, and thus, the number of possible path descriptions is not more than

$$\ell(2d+2)^{2(2d+2)(k^d + d\ell k^{d-1})}.$$

As different marked sets yield different depth first search paths, the latter quantity serves as an upper bound for N_ϵ.

PROOF THAT $\mathcal{G}_\epsilon$ IS AN ϵ-COVER. Let f be a monotone density. On the i-th of the k^d cells of $[0,1]^d$, we define the integers $s_i \leq r_i$, where $Bs_i/\ell \leq f(x) \leq Br_i/\ell$ on that cell, s_i is minimal, and r_i is maximal. It should be clear that the skeleton has at least one member g satisfying the same bounds for all cells i jointly (just verify that the integral is one by adjusting the values of g on the cells without losing monotonicity). Mark all cells visited by f. Clearly, $\int |f-g|$ is not more than the total volume of these cells. But it is not hard to see that the collection of these marked cells is of the form handled by the depth first search counting argument above. In particular, there cannot be more than $(2d+2)(k^d+d\ell k^{d-1})$ such cells. Thus,

$$\int |f-g| \le \frac{B(2d+2)(k^d+d\ell k^{d-1})}{\ell k^d} \le \frac{B(2d+2)}{\ell} + \frac{Bd(2d+2)}{k}.$$

We have an ϵ-cover if

$$k = d\ell,\ \ell = \left\lceil \frac{B(4d+4)}{\epsilon} \right\rceil .$$

For $\epsilon \le 1$, $\ell \le (4d+5)B/\epsilon$.

CONCLUSION OF THE PROOF. Finally, using a trivial bound on the logarithm,

$$\begin{aligned}
\log_{2d+2} N_\epsilon &\le \log_{2d+2} \ell + (4d+4)k^d + (4d+4)d\ell k^{d-1} \\
&= \log_{2d+2} \ell + 8(d+1)k^d \\
&\le \log_{2d+2} \left(\frac{B(4d+5)}{\epsilon}\right) + 8(d+1)d^d \left(\frac{B(4d+5)}{\epsilon}\right)^d \\
&\le \left(1+8(d+1)d^d\right)\left(\frac{B(4d+5)}{\epsilon}\right)^d \\
&\stackrel{\text{def}}{=} \left(\frac{CB^d}{\epsilon}\right)^d
\end{aligned}$$

for some constant C only depending upon d. □

The reader should contrast this entropy, which is polynomial in $1/\epsilon$, with that of the previous section. This polynomial behavior is typical of large (but not huge) nonparametric classes. The bound of Theorem 1 is optimized when ϵ decreases as $B^{d/(d+2)}/n^{1/(d+2)}$, and thus,

$$\sup_{f\in\mathcal{F}} \mathbf{E}\left\{\int |f_n - f|\right\} \le CB^{d/(d+2)}/n^{1/(d+2)}$$

for some constant C depending upon d only. It is known from minimax theory for $d = 1$ that no estimate can have a better expected error rate than $(\log(B)/n)^{1/3}$, so the skeleton estimate has the correct dependence upon n but not upon B.

§7.6. How To Make Gigantic Totally Bounded Classes

There are numerous ways to create classes with Kolmogorov entropy growing exponentially or even double exponentially in $1/\epsilon$. These classes are in fact way past the standard range for nonparametric classes, of which an example was given in the previous section. For these classes, it is impossible to construct density estimates that converge fast uniformly over all densities in the class. Let us give a construction that will lead to some surprising new results.

Given is a positive symmetric unimodal integrable function g on the real line, with total integral $G > 1$. It is assumed that g decreases strictly for $x > 0$. Let $F(x) = \int_x^\infty g(u)\,du$ describe the tail mass. F^{inv} denotes the inverse of F. Let $\mathcal{U}$ denote the class of all unimodal densities f on the real line with a mode at 0, satisfying $f \leq g$. Examples of such classes abound. See the exercises where such bounds are derived for two classes:

A. The class of log-concave (hence unimodal) densities, with a mode at 0, and scaled such that $f(0) = 1$.
B. The class of symmetric unimodal densities with finite absolute r-th moment not exceeding $m < \infty$.

For $\mathcal{U}$, we offer a proof of the following lemma. The implications of this for the two examples above are given after the lemma. For the details, see the exercises.

Lemma 7.2. *For* $\mathcal{U}$:

$$\log_2 N_\epsilon \leq \log_2(g(0)\ell) + 2g(0)\ell + 2t\left\lceil \frac{4g(0)}{\epsilon} \right\rceil,$$

where ℓ is an integer so large that

$$2\int_0^\infty \min\left(\frac{1}{\ell}, g(x)\right)\,dx \leq \frac{\epsilon}{2}$$

and $t > 0$ is so large that $g(t) \leq 1/\ell$.

Proof. We construct a skeleton set by gridding the space $\mathbf{R} \times [0,\infty)$ by cells of dimensions $1/k \times 1/\ell$, with k, ℓ integers to be picked further on. Our skeleton densities take integer multiple values of $1/\ell$ on each x-interval $[j/k, (j+1)/k)$, $j \in \mathcal{Z}$. Also, they must all be tucked under g. Let $t = \sup\{x : g(x) \geq 1/\ell\}$. Clearly, every skeleton density is zero outside $[-t, t]$. By looking at the trace of the graph of a skeleton density starting from the position at the origin without lifting the pen, we note that we must draw at most tk horizontal segments, and at most $g(0)\ell$ vertical edges, and at each grid crossing, we only have two options, to go horizontal to the right, or vertical down. The number of paths traced in this manner is thus at most $g(0)\ell$ (for picking the value at the origin) times $2^{2tk+2g(0)\ell}$, where the 2 in the exponent comes from considering the path on the negative real axis as well. Thus,

$$N_\epsilon \le g(0)\ell 2^{2tk+2g(0)\ell}.$$

Next, we verify that we have an ϵ-cover by the proper choice of k and ℓ. This will show at once that the class is totally bounded, as well. Consider first the maximal unimodal skeleton function $h \le f$, where $f \in \mathcal{U}$. This h does not usually integrate to one and is not a skeleton density. Now, start with intervals $[0/k, 1/k)$ and $[-1/k, 0/k)$, and increase the value of h by $1/\ell$ if allowed (we cannot exceed g), then consider $[1/k, 2/k)$, $[-2/k, -1/k)$, and so forth, until the integral of h is one. For k and ℓ large enough, this procedure will halt. Then the L_1 distance between the skeleton density h and f is at most $2t/\ell + 2g(0)/k + 2\int_t^\infty g$, by simple considerations involving marking grid cells crossed by f. This in turn is not more than $2\int_0^\infty \min(1/\ell, g(x))\,dx + 2g(0)/k$. To get an ϵ-cover, make each term less than $\epsilon/2$. Thus, we pick $k = \lceil 4g(0)/\epsilon \rceil$. □

For the standardized log-concave class, Lemma 2 implies that

$$\log_2 N_\epsilon \le (18e + o(1)) \frac{1}{\epsilon} \log \frac{1}{\epsilon}$$

as $\epsilon \to 0$. Minimizing the bound of Theorem 1 yields the result

$$\sup_{f \in \mathcal{U}} \mathbf{E}\left\{ \int |f_n - f| \right\} = O\left(\left(\frac{\log n}{n} \right)^{1/3} \right),$$

where f_n is a skeleton estimate with ϵ picked appropriately as a function of n (see the exercises). For log-concave densities in particular, this is not best possible, but for all standardized unimodal densities with exponential tails, this is a good bound. We refer to the exercises for further examples, and in particular for the construction of a class in which g is slowly decreasing (but integrable), and the Kolmogorov entropy grows exponentially in $1/\epsilon$, leading to an extremely slow error rate for the skeleton estimate:

$$\sup_{f \in \mathcal{U}} \mathbf{E}\left\{ \int |f_n - f| \right\} = O\left(\frac{1}{\log n} \right).$$

In contrast, for standard nonparametric classes such as the example of the previous section, $\log_2 N_\epsilon$ grows only polynomially in $1/\epsilon$ as $\epsilon \to 0$, which leads in turn to error estimates for the L_1 error that are polynomially decreasing in n.

§7.7. Bibliographic Remarks

The idea of using minimal ϵ-covers of spaces of functions originated in Kolmogorov and Tikhomirov (1961). Yatracos (1985) built on this idea, and had the insight to pick the right classes of sets for use with L_1 distances between densities. Most of this chapter is just a rehash of his work in that paper. Further examples and discussions may be found in Devroye (1987), Yatracos (1988), and Khasminskii and Ibragimov (1990). More examples of entropy calculations are provided by Lorentz (1966) and Makovoz (1986). The one-dimensional uniform rates of convergence over bounded monotone densities on $[0,1]$ were obtained by Birgé (1987a,b, 1989) (see also Chapter 8 of Devroye, 1987). The necessary and sufficient condition for total boundedness suggested in an exercise is due to Kolmogorov (see also Yosida, 1980). Finally, the inequalities for log-concave and unimodal finite moment classes are from Devroye (1984).

§7.8. Exercises

Exercise 7.1 (FINITE MIXTURES). Let $\mathcal{F}$ be the class of densities that are convex combinations of two fixed densities g_1, g_2 on $\mathbf{R}^d$, that is, all densities of the form $pg_1 + (1-p)g_2$, where $p \in [0,1]$. Show how you can find a skeleton density estimate f_n (based upon n i.i.d. points drawn from the unknown mixture) such that

$$\sup_{f\in\mathcal{F}} \mathbf{E}\left\{\int |f_n - f|\right\} \le \frac{C}{\sqrt{n}}$$

for some constant C. NOTE: Estimating p is futile, as it may not be uniquely defined (e.g., when $g_1 = g_2$). You might want to try the maximum likelihood method, but this may cause problems when g_1 or g_2 are unbounded.

Exercise 7.2 (TOTAL BOUNDEDNESS). Show that a collection $\mathcal{F}$ of densities on the real line is totally bounded if and only if these conditions hold:

$$\lim_{h\to 0} \sup_{f\in\mathcal{F}} \int |f(x+h) - f(x)|\, dx = 0\ ;$$

$$\lim_{M\to\infty} \sup_{f\in\mathcal{F}} \int_{|x|>M} |f(x)|\, dx = 0.$$

Exercise 7.3 (SMOOTHNESS CLASSES). Consider the classes $\mathcal{W}^r$ and $\mathcal{W}^r_\infty$ defined for integer $r \ge 1$ as follows: $\mathcal{W}^r$ has densities f on $[0,1]$ that have $r-1$ absolutely continuous derivatives and $\int_0^1 |f^{(r)}| \le C$ for a fixed given C. For example, $\mathcal{W}^1$ contains all absolutely continuous densities on $[0,1]$ with $\int |f'| \le C$. Clearly, then, $f \le C$ as well. For

$\mathcal{W}^r_\infty$, add the condition that $\sup_{[0,1]} |f^{(r)}| \leq C$. Show that the Kolmogorov entropy of $\mathcal{W}^r$ is $O(1/\epsilon^r)$. HINT: Try the case $r = 1$ first, and use the tiling method employed in the text for monotone densities.

Exercise 7.4 (LIPSCHITZ DENSITIES). Obtain matching upper and lower bounds for $\log_2 N_\epsilon$ for all Lipschitz densities on $[0,1]$ with given Lipschitz constant C. The ratio of upper to lower bound should thus be $O(1)$ as $\epsilon \to 0$.

Exercise 7.5 How fast does the Kolmogorov entropy grow for the class of all uniform densities on convex sets of unit volume that are contained in $[-1,1]^d$?

Exercise 7.6 (LOG-CONCAVE DENSITIES). Let $\mathcal{L}$ be the class of densities f on the real line that are log-concave (i.e., $\log f$ is concave), have a mode at zero (note: all log-concave densities are unimodal, although the mode may not be unique), and are rescaled to have $f(0) = 1$.

A. Show that $f(x) \leq e^{1-|x|}$, $x \in \mathbf{R}$.
B. Let $\log_2 N_\epsilon$ be the Kolmogorov entropy of the class $\mathcal{E}$ of all unimodal densities f with mode at the origin that satisfy $f(x) \leq e^{1-|x|}$. Show that for $\mathcal{E}$:

$$\log_2 N_\epsilon \leq (18e + o(1))\frac{1}{\epsilon}\log\frac{1}{\epsilon}$$

where $o(1)$ is with respect to $\epsilon \to 0$. HINT: Apply Lemma 2 and show that

$$2\int_0^\infty \min\left(\frac{1}{\ell}, g(x)\right)\, dx = \frac{4 + 2\log \ell}{\ell}.$$

C. For $\mathcal{E}$, show that we can choose skeletons as a function of n such that for the skeleton estimate f_n:

$$\sup_{f\in\mathcal{E}} \mathbf{E}\left\{\int |f_n - f|\right\} = O\left(\left(\frac{\log n}{n}\right)^{1/3}\right).$$

Exercise 7.7 (UNIMODAL FINITE MOMENT CLASSES). Let $\mathcal{M}$ be the class of symmetric unimodal densities on the real line with finite r-th absolute moment $\int |x|^r f(x)\, dx \leq m < \infty$.

A. Show that $f(x) \leq \min(f(0), m(r+1)/(2|x|^{r+1}))$, $x \in \mathbf{R}$.
B. Derive an upper bound for the Kolmogorov entropy of $\mathcal{M}$.
C. Derive a good upper bound for $\sup_{f\in\mathcal{M}} \mathbf{E}\{\int |f_n - f|\}$, where f_n is a skeleton estimate for $\mathcal{M}$, in which ϵ is chosen by you as a function of n.

Exercise 7.8 (LARGE KOLMOGOROV ENTROPY CLASSES). Let $\mathcal{H}$ be the class of all unimodal densities f on the real line with a mode at 0, such that for all x, $f(x) \leq g(x) = 1/((|x|+1)\log^2(|x|+1))$. Show that the Kolmogorov entropy for this class is $O(\exp(2/\epsilon))$ as $\epsilon \to 0$, and that as a consequence, there exists a skeleton estimate f_n for $\mathcal{H}$ with ϵ chosen as a function of n in such a way that

$$\sup_{f \in \mathcal{H}} \mathbf{E}\left\{\int |f_n - f|\right\} = O\left(\frac{1}{\log n}\right).$$

Obtain a lower bound for $\log_2 N_\epsilon$ that is as close as possible to the upper bound.

§7.9. References

L. Birgé, "On the risk of histograms for estimating decreasing densities," *Annals of Statistics*, vol. 15, pp. 1013–1022, 1987a.

L. Birgé, "Estimating a density under order restrictions: Nonasymptotic minimax risk," *Annals of Statistics*, vol. 15, pp. 995–1012, 1987b.

L. Birgé, "The Grenander estimator: A nonasymptotic approach," *Annals of Statistics*, vol. 17, pp. 1532–1549, 1989.

L. Devroye, "A simple algorithm for generating random variates with a log-concave density," *Computing*, vol. 33, pp. 247–257, 1984.

L. Devroye, *A Course in Density Estimation*, Birkhäuser-Verlag, Boston, 1987.

R. Khasminskii and I. Ibragimov, "On density estimation in the view of Kolmogorov's ideas in approximation theory," *Annals of Statistics*, vol. 18, pp. 999–1010, 1990.

A. N. Kolmogorov and V. M. Tikhomirov, "ϵ-entropy and ϵ-capacity of sets in function spaces," *Translations of the American Mathematical Society*, vol. 17, pp. 277–364, 1961.

G. G. Lorentz, *Approximation of Functions*, Holt, Rinehart and Winston, New York, 1966.

Y. Makovoz, "On the Kolmogorov complexity of functions of finite smoothness," *Journal of Complexity*, vol. 2, pp. 121–130, 1986.

R. E. Tarjan, *Data Structures and Network Algorithms*, CBMS 44, Society for Industrial and Applied Mathematics, Philadelphia, PA, 1983.

Y. G. Yatracos, "Rates of convergence of minimum distance estimators and Kolmogorov's entropy," *Annals of Statistics*, vol. 13, pp. 768–774, 1985.

Y. G. Yatracos, "A note on L_1 consistent estimation," *Canadian Journal of Statistics*, vol. 16, pp. 283–292, 1988.

K. Yosida, *Functional Analysis*, Springer-Verlag, Berlin, 1980.

Chapter 8
The Minimum Distance Estimate: Examples

§8.1. Problem Formulation

Given a class $\mathcal{F}$ of densities, we are to construct a density estimate such that it performs almost as well as the best density in the class. We saw that the skeleton estimate defined in the previous chapter always works when $\mathcal{F}$ is totally bounded. In this chapter we analyze the minimum distance estimate described in Section 6.8. Assume that the densities $f_\theta \in \mathcal{F}$ are indexed by a parameter $\theta \in \Theta$. Recall that the minimum distance estimate selects a density $f_n = f_\theta$ from $\mathcal{F}$ satisfying $\Delta_\theta < \inf_{\theta' \in \Theta} \Delta_{\theta'} + 1/n$, where for each $\theta \in \Theta$:

$$\Delta_\theta = \sup_{A \in \mathcal{A}} \left| \int_A f_\theta - \mu_n(A) \right|.$$

Here the Yatracos class $\mathcal{A}$ is the collection of all sets of the form $\{x : f_\theta(x) > f_{\theta'}(x)\}$, $\theta \neq \theta'$, and μ_n is the standard empirical measure based on the sample $X_1, \ldots, X_n$. Then Theorem 6.4 asserts that for all f:

$$\int |f_n - f| \leq 3 \inf_{\theta \in \Theta} \int |f_\theta - f| + 4 \sup_{A \in \mathcal{A}} \left| \int_A f - \mu_n(A) \right| + \frac{3}{n}.$$

In particular, if $f \in \mathcal{F}$, then

$$\int |f_n - f| \leq 4 \sup_{A \in \mathcal{A}} \left| \int_A f - \mu_n(A) \right| + \frac{3}{n}.$$

The right-hand side of the last inequality is of the order of $n^{-1/2}$ whenever $\mathcal{A}$ has a finite Vapnik–Chervonenkis dimension. The problem from this point on is purely combinatorial. In the rest of this chapter we work out examples for different classes.

§8.2. Series Estimates

Let $\psi_1, \ldots, \psi_k$ be fixed basis functions from $\mathbf{R}^d$ to $\mathbf{R}$ such that $\int \psi_i = r_i$ for all $1 \le i \le k$. We define the class $\mathcal{F}_k$ as the class of all linear combinations of the basis functions $f_\theta = \sum_{i=1}^k a_i \psi_i$ with coefficient vector $\theta = (a_1, \ldots, a_k)$ satisfying $\sum_{i=1}^k a_i r_i = 1$. The latter condition assures that all candidates f_θ have integral equal to one, a necessary condition to apply Theorem 6.4 for the performance of the minimum distance estimate. (Recall from Exercise 6.2 that no restriction of the parameters is necessary if one is willing to give up a constant factor in the performance bound.) In this case, all sets $A_{\theta,\theta'} = \{x : f_\theta(x) > f_{\theta'}(x)\}$ are of the form

$$\left\{ x : \sum_{i=1}^k b_i \psi_i(x) > 0 \right\},$$

where $\theta' = (a'_1, \ldots, a'_k)$, and $b_i = a_i - a'_i$. Lemma 4.2 shows that the Vapnik–Chervonenkis dimension of $\mathcal{A}$ is at most k, so invoking Theorem 4.4, we obtain, for the minimum distance estimate f_n:

$$\mathbf{E}\left\{ \int |f_n - f| \right\} \le 3 \inf_{g \in \mathcal{F}_k} \int |g - f| + c\sqrt{\frac{k}{n}} + \frac{3}{n},$$

where c is a universal constant. It is important to emphasize the nonasymptotic character of the above bound. In fact, k may be varied with n, trying to obtain an optimal balance between the error of the best estimator in $\mathcal{F}_k$—which may be thought of as a bias—and the term $c\sqrt{k/n}$, which plays the role of the variation. Of course, the size of the first term may be very different for different densities, but for certain "smooth" densities, standard approximation-theoretic results may be applied to obtain bounds.

L_2-based theory suggests picking the coefficient of the ith basis function as

$$a_{ni} = \frac{1}{n} \sum_{j=1}^n \psi_i(X_j).$$

Indeed, if the basis functions are orthonormal (i.e., $\int \psi_i^2 = 1$ and $\int \psi_i \psi_j = 0$ for all $1 \le i, j \le k$, $i \ne j$), and $f \in L_2$, then a_{ni} is the natural estimate of the Fourier coefficient $a_i^* = \int \psi_i f$. However, simplicity has a price: the series estimate with the L_2-based coefficients may be much inferior to the one based on optimally chosen parameters. We illustrate this in the following very simple example.

Assume that f is a density supported on $[-\pi, \pi]$. Consider the basis functions $\psi_1, \ldots, \psi_k$ as the first k of the standard trigonometric system

$$\psi_1(x) \equiv \frac{1}{\sqrt{2\pi}}, \ \psi_{2i}(x) = \frac{\cos(ix)}{\sqrt{\pi}}, \ \psi_{2i+1}(x) = \frac{\sin(ix)}{\sqrt{\pi}}, \ i \ge 1.$$

It is well known by the Weierstrass approximation theorem that $\bigcup_{k=1}^{\infty} \mathcal{F}_k$ is dense (with respect to the L_1 distance) in the class of all densities supported on $[-\pi, \pi]$, which, together with the performance bound above, immediately implies that for any such density, the error $\mathbf{E}\{\int |f_n - f|\}$ of the minimum distance estimate goes to zero whenever $k \to \infty$ such that $k/n \to 0$. In contrast, it is known that for some densities the error of the series estimate with the L_2-based coefficients can converge to infinity at any rate that is $o(\log k)$. To see how bad this is, just note that for any estimate f_n which is a density (i.e., nonnegative and $\int f_n = 1$), the error is bounded by two.

§8.3. Parametric Estimates: Exponential Families

Often one suspects that the density f belongs to some small class. In such cases parametric estimates may be appealing. An eternal problem is robustness: What happens if the parametric model is false? The minimum distance estimate guarantees that as long as the true density is not far from the hypothesized class, the error of the estimate does not blow up either.

As a warm-up exercise, assume that the data are believed to be drawn from a normal density on the real line. This leads us to the construction of the class

$$\mathcal{F} = \left\{ f_{m,\sigma}(x) = \frac{1}{\sigma\sqrt{2\pi}} e^{-(x-m)^2/2\sigma^2} : m \in \mathbf{R}, \sigma > 0 \right\}.$$

To determine the performance of the minimum distance estimate, we must estimate the Vapnik–Chervonenkis dimension of the corresponding Yatracos class $\mathcal{A}$. Note that every set in $\mathcal{A}$ may be written in the form of $\{x : ax^2 + bx + c \geq 0\}$ for some coefficients $a, b, c \in \mathbf{R}$, that is, $\mathcal{A}$ contains either closed intervals, or the union of two closed half-infinite intervals. The Vapnik–Chervonenkis dimension of this class is easily seen to be three.

The argument above may be significantly generalized as follows. A family $\mathcal{F}$ of densities on $\mathbf{R}^d$ is called an *exponential family* if each density in $\mathcal{F}$ may be written in the form

$$f_\theta(x) = c\alpha(\theta)\beta(x)e^{\sum_{i=1}^{k} \pi_i(\theta)\psi_i(x)},$$

where θ belongs to some parameter set Θ, $\psi_1, \ldots, \psi_k : \mathbf{R}^d \to \mathbf{R}$, $\beta : \mathbf{R}^d \to [0, \infty)$, $\alpha, \pi_1, \ldots, \pi_k : \Theta \to \mathbf{R}$ are fixed functions, and c is a normalizing constant. Examples of exponential families include classes of Gaussian, gamma, beta, Rayleigh, and Maxwell densities.

THEOREM 8.1. *Let $\mathcal{F}$ be an exponential family of densities defined as above, and let f_n be the minimum distance estimate based on this class. Then for any $f \in \mathcal{F}$:*

$$\mathbf{E}\left\{\int |f_n - f|\right\} \leq c\sqrt{\frac{k+1}{n}} + \frac{3}{n},$$

where c is a universal constant. Moreover, for any density f:

$$\mathbf{E}\left\{\int |f_n - f|\right\} \leq 3 \inf_{\theta\in\Theta} \int |f_\theta - f| + c\sqrt{\frac{k+1}{n}} + \frac{3}{n}.$$

PROOF. Note that $f_\theta(x) > f_{\theta'}(x)$ if and only if

$$\sum_{i=1}^{k}(\pi_i(\theta) - \pi_i(\theta'))\psi_i(x) + \log\frac{\alpha(\theta)}{\alpha(\theta')} > 0.$$

The class of functions appearing on the left-hand side spans a $(k+1)$-dimensional vector space, and so Lemma 4.2 implies that the Vapnik–Chervonenkis dimension of $\mathcal{A}$ is at most $k+1$. □

§8.4. Neural Network Estimates

A feed-forward one-hidden-layer *neural network* (or simply neural network) with k hidden nodes is a function $\phi : \mathbf{R}^d \to \mathbf{R}$ of the form

$$\phi(x) = \sum_{i=1}^{k} a_i\sigma(b_i^T x + c_i),$$

where $\sigma : \mathbf{R} \to \mathbf{R}$ is a fixed *activation function*, $a_i, c_i \in \mathbf{R}$, and $b_i \in \mathbf{R}^d$, $i = 1, \ldots, k$, are the parameters of the network, and $b_i^T x$ denotes the usual inner product of the vectors b_i and $x \in \mathbf{R}^d$. In order to generate a class eligible for the minimum distance method, we must require that $\int \sigma < \infty$. Define $\mathcal{F}$ as the set of neural networks ϕ. Once again, to obtain a performance bound for the minimum distance density estimate, we need upper bounds for the Vapnik–Chervonenkis dimension of the associated Yatracos class. Each set in the class is of the form

$$\left\{x : \sum_{i=1}^{k}\left(a_i\sigma(b_i^T x + c_i) - a_i'\sigma(b_i'^T x + c_i')\right) > 0\right\}.$$

Observe that $\mathcal{A}$ is a subclass of the class of all sets of the form $\{x : \psi(x) > 0\}$, where ψ is a neural network of $2k$ hidden nodes. Computation of the Vapnik–Chervonenkis dimension of such classes of sets has been the subject of intensive research in pattern recognition and learning theory. It is known that the Vapnik–Chervonenkis dimension may be finite or infinite, depending on the activation function σ. The monograph of Anthony and Bartlett (1999) contains, among other things, the following useful result.

THEOREM 8.2 (ANTHONY AND BARTLETT, 1999). *Let $\Theta \subset \mathbf{R}^m$ be a parameter set, and consider a class of functions $g_\theta : \mathbf{R}^d \to \mathbf{R}$ parametrized by Θ. Assume that for each $\theta \in \Theta$ and $x \in \mathbf{R}^d$, $g_\theta(x)$ may be computed by at most t steps of an algorithm which in each step executes one of the following operations: the arithmetic operations $+, -, \cdot$, and $/$ on real numbers; the exponential function e^x on real numbers; an indicator function of the form $\mathbf{1}_{[x>a]}$, $\mathbf{1}_{[x\geq a]}$, $\mathbf{1}_{[x<a]}$, $\mathbf{1}_{[x\leq a]}$, $\mathbf{1}_{[x=a]}$, $\mathbf{1}_{[x\neq a]}$, for $a \in \mathbf{R}$. Then the Vapnik–Chervonenkis dimension V of the class of sets $\{\{x : g_\theta > 0\} : \theta \in \Theta\}$ is at most*

$$V \leq t^2 m(m + 19\log_2(9m)).$$

If the exponential function is evaluated at most q times for each pair (x, θ), then

$$V \leq m^2(q+1)^2 + 11m(q+1)(t + \log_2(9m(q+1))).$$

The above theorem guarantees the finiteness of the Vapnik–Chervonenkis dimension of the neural network class for a whole host of all activation functions, though in many cases it does not provide the best possible bound. Of course, Theorem 2 may be used in many other situations, not just for neural network classes.

§8.5. Mixture Classes, Radial Basis Function Networks

Consider first the class of all mixtures of k normal densities in $\mathbf{R}$, that is, the class of all densities of form

$$\phi(x) = \frac{1}{\sqrt{2\pi}} \sum_{i=1}^{k} \frac{p_i}{\sigma_i} e^{-(x-m_i)^2/2\sigma_i^2},$$

where $(p_1, \ldots, p_k)$ is a probability vector, $\sigma_1, \ldots, \sigma_k$ are positive numbers, and $m_1, \ldots, m_k$ are arbitrary real numbers. The corresponding Yatracos class contains all sets of the form

$$\{x : \phi(x) > \phi'(x)\}$$
$$= \left\{ x : \sum_{i=1}^{k} \left(\frac{p_i}{\sigma_i} e^{-(x-m_i)^2/2\sigma_i^2} - \frac{p_i'}{\sigma_i'} e^{-(x-m_i')^2/2\sigma_i'^2} \right) > 0 \right\}.$$

This class of sets may be written in the form given in Theorem 2 with $m = 6k$, where the parameter set Θ is a subset of the $6k$-dimensional vector space of the parameters $p_i, p'_i, m_i, m'_i, \sigma_i, \sigma'_i$, $i = 1, \ldots, k$. Simple counting reveals that given x and the vector of parameters, each function may be computed with no more than $t = 14k$ operations admitted by Theorem 2. Hence, we obtain that the Vapnik–Chervonenkis dimension of the Yatracos class based on all mixtures of k univariate normal densities is bounded by a constant times k^4. We suspect that this estimate is loose. Nevertheless, the finiteness of the Vapnik–Chervonenkis dimension implies that the error of the minimum distance estimate is not more than a constant times $k^2/\sqrt{n}$, whenever the true density f is, in fact, a mixture of k normals.

The same argument may be generalized to the multivariate setting. Consider now the class $\mathcal{F}$ of all mixtures of k normal densities over $\mathbf{R}^d$:

$$\phi(x) = \sum_{i=1}^{k} \frac{p_i}{\sqrt{(2\pi)^d \det(\Sigma_i)}} e^{-\frac{1}{2}(x-m_i)^T \Sigma_i^{-1}(x-m)},$$

where $(p_1, \ldots, p_k)$ is a probability vector, $\Sigma_k, \ldots, \Sigma_k$ are positive definite $d \times d$ matrices, and $m_1, \ldots, m_k$ are arbitrary elements of $\mathbf{R}^d$. Then it is clear from Theorem 2 that the Vapnik–Chervonenkis dimension of the corresponding Yatracos class is finite, and it is a matter of straightforward counting to obtain an explicit upper bound. In any case, if f is a mixture of k normal densities, and is estimated by the minimum distance estimate based on the class defined here, then

$$\mathbf{E}\left\{\int |f_n - f|\right\} \leq \sqrt{\frac{c_{k,d}}{n}},$$

where $c_{k,d}$ is a constant depending on k and d.

Often one may obtain much sharper estimates by direct methods. As an illustration, consider the class $\mathcal{F}$ of all mixtures of k exponential densities (i.e., translations and scales of $e^{-x}, x \geq 0$). Again, it suffices to bound the Vapnik–Chervonenkis dimension of

$$\mathcal{A} = \{x : f(x) > g(x); f, g \in \mathcal{F}\}.$$

A member set in this class is thus of the form

$$\left\{x : \sum_{i=1}^{2k} a_i e^{-b_i x} I_{[x > c_i]} > 0\right\},$$

where $a_i, c_i \in \mathbf{R}$ and $b_i > 0$ are free parameters. Now clearly, we have at most $k+1$ intervals defined by the thresholds c_i. On the other hand, on each of these intervals, a set of the form $\{x : \sum_{i=1}^{2k} a_i e^{-b_i x} > 0\}$ defines at most $2k+1$ intervals, by Lemma 4.5. Therefore, each set in the class $\mathcal{A}$ is the union of at most $(2k+1)(k+1)$ intervals, and the Vapnik–Chervonenkis dimension of $\mathcal{A}$ is not more than $2(2k+1)(k+

$1) = O(k^2)$. Contrast this with the $O(k^4)$ bound obtainable by Theorem 2.

§8.6. Bibliographic Remarks

For a survey on orthogonal series estimates and their L_1 aspects, we refer to Chapter 12 of Devroye and Györfi (1985). Surveys of approximation properties of orthogonal series may be found in Butzer and Nessel (1971). A good account of the Weierstrass approximation theorem may be found there. Theorem 2 and related results appear in Anthony and Bartlett (1999), who build on work of Goldberg and Jerrum (1995), Khovanskii (1991), Karpinsky and Macintyre (1997), Macintyre and Sontag (1993), and Koiran and Sontag (1997). Several related results are also surveyed in Devroye, Györfi, and Lugosi (1996). Radial basis function networks originated in the work of Poggio and Girosi (1990) and were analyzed in a series of subsequent papers, such as in Krzyżak and Linder (1998).

§8.7. Exercises

Exercise 8.1 Use Theorem 2 to obtain an explicit upper bound for the Vapnik–Chervonenkis dimension of the Yatracos class based on the family of all mixtures of k normal densities in $\mathbf{R}^d$. HINT: In order to avoid having to deal with matrix inversions, you might consider the inverse covariance matrices as parts of the parameter vector θ.

Exercise 8.2 (TRIGONOMETRIC SERIES ESTIMATES). Let f be a density on $[-\pi, \pi]$. Let $S_m(f)$ be the trigonometric series $\sum_{i=1}^m a_i \psi_i(x)$ truncated at m terms, where the ψ_i's are as in the text, and the a_i's are weights. If $a_i = \int f \psi_i$, then it is known that $\lim_{m \to \infty} \int |S_m(f) - f| = 0$ whenever $\int f \log(f+1) < \infty$. In the trigonometric series estimate, m is a function of n such that $m \to \infty$ and $m/n \to 0$, $f_n = \sum_{i=1}^m a_{ni} \psi_i(x)$, and $a_{ni} = (1/n) \sum_{j=1}^n \psi_i(X_j)$.

A. Show that $\mathbf{E}\{f_n(x)\} = S_m(f)(x)$. Show that $\mathbf{E}\{\int |f_n - S_m(f)|\} \le \sqrt{m/n}$.
B. If $\int f^2 < \infty$, show that $\int |S_m(f) - f| \le \sqrt{2\pi \sum_{i>m} a_i^2}$.
C. Show that $\mathbf{E}\{\int |f_n - f|\} \to 0$ when $\int f^2 < \infty$.
D. When f is Lipschitz with constant C on $[-\pi, \pi]$, show that $\int |S_m(f) - f| \le aC/m$ for some constant a. Conclude that $\mathbf{E}\{\int |f_n - f|\} = O(1/n^{1/3})$.
E. Find a density f on $[-\pi, \pi]$ with the property that $a_m \log m \to \infty$ as $m \to \infty$.
F. For the density of Part E, show that $\lim_{n \to \infty} \mathbf{E}\{\int |f_n - f|\} = \infty$.

Exercise 8.3 (BERNSTEIN POLYNOMIALS0. Let f be a density on $[0,1]$. The Bernstein polynomials of order m are all polynomials of the form $x^a(1-x)^b$, with integer-valued $a, b \geq 0$ such that $a + b = m$. Let $\mathcal{F}_m$ be the linear vector space spanned by these functions. We define the minimum distance Bernstein density estimate f_n as the minimum distance estimate based on $\mathcal{F}_m$.

A. Show that for all densities f on $[0,1]$, the minimum distance Bernstein estimate is consistent: $\lim_{n\to\infty} \mathbf{E}\{\int |f_n - f|\} \to 0$.
B. Compute the Vapnik–Chervonenkis dimension of the Yatracos class induced by $\mathcal{F}_m$.
C. Based on an estimate similar to the minimum distance Bernstein estimate, suggest a method that outputs a random variable $Z \in \{0,1\}$, where $Z = 1$ indicates a belief that f is a polynomial on $[0,1]$ of degree 20 or less, and $Z = 0$ is its complement. Let $Y(f)$ be the indicator of f being such a polynomial. Design a method Z such that for every f, $\lim_{n\to\infty} \mathbf{P}\{Z = Y(f)\} = 1$.
D. Repeat Part C to test whether f is a finite polynomial (of an arbitrary degree).

§8.8. References

M. Anthony and P. L. Bartlett, *Neural Network Learning: Theoretical Foundations*, Cambridge University Press, Cambridge, 1999.

P. L. Butzer and R. J. Nessel, *Fourier Analysis and Approximation*, Birkhäuser-Verlag, Basel, 1971.

L. Devroye and L. Györfi, *Nonparametric Density Estimation: The L_1 View*, Wiley, New York, 1985.

L. Devroye, L. Györfi, and G. Lugosi, *A Probabilistic Theory of Pattern Recognition*, Springer-Verlag, New York, 1996.

P. Goldberg and M. Jerrum, "Bounding the Vapnik-Chervonenkis dimension of concept classes parametrized by real numbers," *Machine Learning*, vol. 18, pp. 131–148, 1995.

M. Karpinski and A. Macintyre, "Polynomial bounds for VC dimension of sigmoidal and general pfaffian neural networks," *Journal of Computer and System Science*, vol. 54, pp. 169–176, 1997.

A. G. Khovanskii, "Fewnomials," in: *Translations of Mathematical Monographs*, vol. 88, American Mathematical Society, Providence, RI, 1991.

P. Koiran and E. D. Sontag, "Neural networks with quadratic VC dimension," *Journal of Computer and System Science*, vol. 54, pp. 190–198, 1997.

A. Krzyżak and T. Linder, "Radial basis function networks and complexity regularization in function learning," *IEEE Transactions on Neural Networks*, vol. 9, pp. 247–256, 1998.

A. Macintyre and E. D. Sontag, "Finiteness results for sigmoidal neural networks," in: *Proceedings of the 25th Annual ACM Symposium on the Theory of Computing*, pp. 325–334, Association of Computing Machinery, New York, 1993.

T. Poggio and F. Girosi, "A theory of networks for approximation and learning," *Proceedings of the IEEE*, vol. 78, pp. 1481–1497, 1990.

Chapter 9
The Kernel Density Estimate

§9.1. Approximating Functions by Convolutions

In this chapter, we get our first taste of real analysis, starting with some results on the approximations of functions in L_1. The literature on this is vast, and a lot of it is ancient. The problem is that f cannot be approximated in L_1 by μ_n, the empirical measure, as the total variation distance between any density f and any atomic measure (like μ_n) is 1. Thus, the approximation itself must have a density. The kernel estimate provides this: it smooths the empirical measure μ_n. This section has no combinatorial contributions, but develops just enough of the function approximation material to understand the remaining chapters.

The convolution $f * g$ of two functions f and g is

$$f * g(x) = \int f(y)g(x-y)\,dy = \int g(y)f(x-y)\,dy,$$

where $x, y \in \mathbf{R}^d$. If f and g are the densities of independent random variables X and Y, then $f * g$ is the density of $X+Y$. It is clear that if Y is concentrated near 0, then $f * g$ should be close to f. Indeed, we have the following fundamental approximation theorem from real analysis.

THEOREM 9.1. *Let K be an arbitrary integrable function on $\mathbf{R}^d$ (i.e., $\int |K| < \infty$), and let f be a density on $\mathbf{R}^d$. Denoting $K_h(x) = (1/h^d)K(x/h)$, $x \in \mathbf{R}^d, h > 0$, we have*

$$\lim_{h \downarrow 0} \int \left| f * K_h - f \int K \right| = 0.$$

PROOF. We may assume without loss of generality that $\int K \in \{0, 1\}$. We will prove the statement when $\int K = 1$, leaving the $\int K = 0$ case as an easy exercise. Assume first that the statement is true for a dense subspace of functions g. Then

$$\begin{aligned}\int |f * K_h - f| &\le \int |f-g| * |K_h| + \int |f-g| + \int |g * K_h - g| \\ &\le \left(\int |K| + 1\right) \int |f-g| + o(1)\end{aligned}$$

as $h \to 0$. Here we made use of Young's inequality (Section 5.5):

$$\int |f * g| \leq \int |f| \cdot \int |g|$$

valid for any integrable functions f, g. The first term on the right-hand side can be made as small as desired by the choice of g and the finiteness of $\int |K|$. So, we only need to prove the theorem for a dense subclass, such as the class of Lipschitz densities of compact support. Thus, let f be Lipschitz with constant C (i.e., $|f(x) - f(y)| \leq C\|x - y\|$, $x, y \in \mathbf{R}^d$), and supported on $[-M, M]^d$ for finite M. Let $L = KI_A$ where I is the indicator function, and $A = [-r, r]^d$ is a large cube of our choice. Then, by Young's inequality,

$$\begin{aligned}
&\int |f * K_h - f| \\
&\quad \leq \int \left| f * L_h - f \int L_h \right| + \int |f| \int |K_h - L_h| + \int |f * (K_h - L_h)| \\
&\quad \leq \int \int |f(x - y) - f(x)||L_h(y)|\, dy\, dx + 2 \int |K - L| \\
&\quad \leq \int_{[-M-r, M+r]^d} \left(\int C\|y\||L_h(y)| dy \right) dx + 2 \int |K - L| \\
&\quad \leq (2M + 2r)^d Crh\sqrt{d} \int |L| + 2 \int |K - L| \\
&\quad \leq (2M + 2r)^d Crh\sqrt{d} \int |K| + 2 \int |K - L| \\
&\quad = o(1) + 2 \int |K - L|,
\end{aligned}$$

which is as small as desired by choice of r (and thus L). □

§9.2. Definition of the Kernel Estimate

If our goal is to find a density estimate for which $\int |f_n - f|$ is small, we might make use of the fact that $\int |f * K_h - f|$ is small when h is small and $\int K = 1$. Functions K with $\int |K| < \infty$ and $\int K = 1$ will be called kernels. Indeed, $f * K_h$ may in turn be approximated by $\mu_n * K_h$, where μ_n is the empirical measure. More explicitly, $\mu_n * K_h$ is

$$f_n(x) = \frac{1}{n} \sum_{i=1}^{n} K_h(x - X_i) = \frac{1}{nh^d} \sum_{i=1}^{n} K\left(\frac{x - X_i}{h}\right)$$

which is nothing but the kernel estimate first proposed and developed by Akaike (1954), Parzen (1962), and Rosenblatt (1956).

Various related estimates have been proposed, and we cite some of them for future reference. The variable kernel estimate is obtained whenever $h = h(x)$ in the above definition. We may no longer have a bona fide density integrating to one, yet the properties of such estimates are not to be discarded right away. We speak of variable kernel estimates of the second kind when $h = h(X_i)$ in the above definition. Note that in this case, $\int f_n = 1$.

For standard kernel estimates as well as variable kernel estimates, we may pick h and K as a function of n and/or the data. This leads to interesting questions that we will ask in later chapters. Unless we mention it explicitly, the data-dependent kernel estimate will be the kernel estimate in which $h = h(X_1, \ldots, X_n)$ is a measurable function of the data: it does not depend upon x as in the variable kernel estimate. However, for each n, a different function is taken. We will not make this dependence explicit, to keep the number of subscripts small.

§9.3. Consistency of the Kernel Estimate

In this section we will show the following theorem, which is valid for all densities f and all kernels K.

THEOREM 9.2. *Let K be a fixed kernel, and let h depend on n only. If $h \to 0$ and $nh^d \to \infty$ as $n \to \infty$, then* $\mathbf{E}\{\int |f_n - f|\} \to 0$.

PROOF. Note that

$$\int |f_n - f| \le \int |f * K_h - f| + \int |\mu_n * K_h - f * K_h|.$$

The first term on the right-hand side, also called the bias term, tends to 0 by Theorem 1. There are two tools that we will use repreatedly in the text. First of all, if ν is any probability measure, and K and L are kernels, then, by Young's inequality, $\int |\nu * K_h - \nu * L_h| \le \int |K - L|$. This inequality applies in particular when ν is f and when it is μ_n. Thus,

$$\int |\mu_n * K_h - f * K_h| \le \int |\mu_n * L_h - f * L_h| + 2 \int |K - L|.$$

The last term can be made as small as desired by choice of L from among bounded kernels with support on a compact hypercube. The second trick uses the fact that $\int \nu * K_h = 1$ for any probability measure ν and kernel K. Thus,

$$\int |\mu_n * L_h - f * L_h| = 2 \int (f * L_h - \mu_n * L_h)_+$$

and therefore

$$\begin{aligned} &\mathbf{E}\left\{\int |\mu_n * L_h - f * L_h|\right\} \\ &\quad = \int \mathbf{E}\{|\mu_n * L_h - f * L_h|\} \end{aligned}$$

$$
\begin{aligned}
&= 2\int \mathbf{E}\{(f * L_h - \mu_n * L_h)_+\} \\
&= 2\int \min\left(f * L_h, \mathbf{E}\{(f * L_h - \mu_n * L_h)_+\}\right) \\
&\le 2\int \min\left(f * L_h, \sqrt{\mathbf{E}\{(f * L_h - \mu_n * L_h)^2\}}\right) \\
&\le 2\int \min\left(f, \sqrt{\mathbf{E}\{(f * L_h - \mu_n * L_h)^2\}}\right) + 2\int |f * L_h - f| \\
&= o(1).
\end{aligned}
$$

The last step follows from $nh^d \to \infty$, the dominated convergence theorem, and Theorem 1. Here we used the fact that $\mu_n * L_h - f * L_h = \sum_{i=1}^n Z_i$, where the Z_i's are i.i.d. zero mean random variables that have variance bounded by $(L_h)^2 * f/n^2$. Thus, $\mathbf{E}\{(\mu_n * L_h - f * L_h)^2\}$ is bounded by $(L^2)_h * f/(nh^d)$, which is $(f \int L^2 + o(1))/(nh^d)$ at all Lebesgue points for x (see Exercise 5.9). □

The remarkable thing about Theorem 2 is that it is valid whenever the kernel $\int K = 1$. For example, if we take K to be the density that is uniformly distributed on the ball of radius 1 centered at $(1000, 0, \ldots, 0)$, then the average $f * K_h$ calculates an integral that does not even include the origin. Furthermore, it is not even necessary that K be a density. Also, there is no possiblity of improving the conditions $h \to 0$ and $nh^d \to \infty$ (see the exercises). In some cases, we have bandwidths that depend upon the data. The proof of the following theorem is left as an exercise.

THEOREM 9.3. *Let K be a fixed kernel. Let the bandwidth H be an arbitrary function of the data such that $H \to 0$ and $nH^d \to \infty$ in probability as $n \to \infty$ (i.e., for every $\epsilon > 0$, $\lim_{n\to\infty} \mathbf{P}\{H + 1/(nH^d) > \epsilon\} = 0$). Then $\mathbf{E}\{\int |f_n - f|\} \to 0$.*

§9.4. Concentration

Consider the quantity $\int |f_n - g|$ where g is an arbitrary function of x and may be substituted by f or $f * K_h$. This quantity is very stable in the sense that if one X_i changes value and the other $n-1$ data points remain fixed (and denoting the new kernel estimate by f_n^*), then $\int |f_n - f_n^*| \le 2\int |K|$. By the bounded difference inequality (Theorem 2.2), we thus have

$$
\mathbf{P}\left\{\left|\int |f_n - g| - \mathbf{E}\int |f_n - g|\right| \ge t\right\} \le 2e^{-t^2/2n(\int |K|)^2}, \ t > 0.
$$

In particular, we have by Chebyshev's inequality:

THEOREM 9.4. *If* $\sqrt{n}\,\mathbf{E}\int|f_n - f| \to \infty$, *then*

$$\frac{\int|f_n - f|}{\mathbf{E}\int|f_n - f|} \to 1 \text{ in probability.}$$

Theorem 4 thus establishes the relative stability of the L_1 error for kernel estimates. It implies that the expected L_1 error is a good measuring stick because the actual error never deviates substantially from it. We will see further on that for all nonnegative kernels, and indeed all kernels that have at least one nonvanishing moment (on the real line), the $\sqrt{n}$ condition is satisfied.

As another by-product, we see that if $\mathbf{E}\int|f_n - f| \to 0$, then $\int|f_n - f| \to 0$ almost surely (by the Borel-Cantelli lemma). Therefore, the equivalence theorem alluded to in the previous section is in fact stronger: $\int|f_n - f| \to 0$ almost surely for all f. Thus, strong and weak convergence of the L_1 error are equivalent.

§9.5. Choosing the Bandwidth

In bandwidth selection, one is interested in functions $H = H(X_1, \ldots, X_n)$ such that $\int|f_{n,H} - f|$ comes close to $\inf_h \int|f_{n,h} - f|$, where $f_{n,h}$ makes the dependence upon h explicit. A bandwidth H with the property that

$$\limsup_{n\to\infty} \frac{\mathbf{E}\{\int|f_{n,H} - f|\}}{\inf_h \mathbf{E}\{\int|f_{n,h} - f|\}} \leq C(f)$$

for some finite $C(f)$ is called a universal bandwidth. It is uniformly universal with constant C if we may replace $C(f)$ by C for all f. This problem has been attacked by many. However, many attempts can be discarded from the outset. Most authors of papers on bandwidth selection try minimizing $\int(f_{n,h} - f)^2$ with respect to h. Even if f were known, and even if f were square integrable, the bandwidth thus selected would not be universally useful. For example, let us denote by H a data-dependent bandwidth for which $\int(f_{n,H} - f)^2 \sim \inf_h \int(f_{n,h} - f)^2$ (so that H is the L_2-optimal choice; for bounded densities, Stone (1984) shows how one can do this; Wegkamp (2000) provides a newer proof). Assume that K is the uniform density on $[-1, 1]$. For the Cauchy density $1/(\pi(1 + x^2))$, $H \sim c/n^{1/5}$ in probability, yet one can check that $\mathbf{E}\{\int|f_{n,H} - f|\}/\inf_h \mathbf{E}\{\int|f_{n,h} - f|\} \to \infty$ as the optimal bandwidth for L_1 is larger than $n^{-1/5}$ (see the exercises). In other words, even within the class of ultra-smooth densities (such as the Cauchy density), minimizing L_2 is just the wrong thing to do—intuitively, squaring tends to squash errors in the tails and make them unimportant.

Others have attempted to pick h by maximum likelihood, for example by maximization of $\prod_{i=1}^{n} f_{n-1,i,h}(X_i)$, where $f_{n-1,i,h}$ is the kernel estimate based on the $n - 1$ data points obtained after deleting X_i. While for densities with compact support on $\mathbf{R}$, the maximizing H is indeed consistent (i.e., $H + 1/(nH^d) \to 0$ in probability), there is again no relationship with the total variation criterion. If the density f has heavier than exponential tails, the method is not even consistent (see the exercises).

Hall and Wand (1988) looked at the asymptotic expansion of $\mathbf{E}\{\int |f_{n,h} - f|\}$ and minimized the main asymptotic terms to obtain a recipe for h as a function of n, f and K. They then estimate the unknown quantity involving f from the data, and propose this as a plug-in bandwidth estimate. For sufficiently smooth and small-tailed densities, and for positive kernels, they were able to show that $\mathbf{E}\{\int |f_{n,H} - f|\} \sim \inf_h \mathbf{E}\{\int |f_{n,h} - f|\}$. However, their method, and all other plug-in methods we are aware of, are not universal bandwidths.

The double kernel method uses a pair of kernels, K and L, and picks $H = \arg\min_h \int |f_{n,h} - g_{n,h}|$, where $f_{n,h}$ and $g_{n,h}$ are the kernel estimates with kernels K and L, respectively. Assume that $d = 1$. If the characteristic functions of K and L do not coincide on an open interval about the origin, then the choice H is consistent (Devroye, 1989a). Furthermore, if K and L are symmetric, bounded kernels of compact support, $K \geq 0$, $\int x^2 L(x)\,dx = 0$, if both K and L are L_1-Lipschitz (i.e., $\int |K_1 - K_h| \leq C(h-1)$ for some $C < \infty$ and all $h > 1$, and similarly for L), and if f is absolutely continuous, and f' is absolutely continuous, and $\int \sqrt{\sup_{|y|\leq 1} f(x+y)}\,dx < \infty$, then

$$\limsup_{n\to\infty} \frac{\mathbf{E}\{\int |f_{n,H} - f|\}}{\inf_h \mathbf{E}\{\int |f_{n,h} - f|\}} \leq \frac{1+\rho}{1-\rho}$$

where $\rho = 4\sqrt{\int L^2 / \int K^2}$ is a constant that can be made as small as desired by choice of L. The behavior outside the smoothness class described above is unknown: for example, we do not know whether the double kernel bandwidth is universal, let alone uniformly universal. The methodology studied in this book was originally meant to fill the hole and it allows us to obtain a uniformly universal bandwidth with constant 3, and in fact, to obtain even stronger nonasymptotic uniformity:

$$\sup_f \frac{\mathbf{E}\{\int |f_{n,H} - f|\}}{\inf_h \mathbf{E}\{\int |f_{n,h} - f|\}} \leq 3 + o(1).$$

Note in particular that this result does not depend on the dimension, and that the bound is uniform over all f.

§9.6. Choosing the Kernel

The selection of the pair (h, K) may be tackled as one problem, and indeed, h may be absorbed into K as a scale factor. Asymptotic evidence for smooth densities shows that for large sample sizes, the shape of the optimal kernel is unique. For example, for $\mathbf{R}^1$, classical L_2 theory (Watson and Leadbetter, 1963) shows that for L_2 errors, among all positive kernels, the Epanechnikov kernel (Epanechnikov, 1969) $K(x) = \max(\frac{3}{4}(1 - x^2), 0)$ is best possible. For $\mathbf{R}^d$, Deheuvels (1977) showed the L_2 optimality of $\max((1 - \|x\|^2)^d, 1)$. For the L_1 error, there is evidence that the Epanechnikov kernel is also best among all positive kernels (see the discussion in the next section). For these reasons, authors have typically decoupled the choices of K and h, and picked K fixed as one of these asymptotically optimal kernels.

We would also like to direct the reader to Chapters 16 and 17, where nonnegative kernels are considered and the issue of the selection of K from this bigger class of functions is revived.

§9.7. Rates of Convergence

The rate of convergence of the L_1 error of the standard kernel estimate is well understood in $\mathbf{R}$ but much less so in $\mathbf{R}^d$. For simplicity, we stick to $\mathbf{R}$ in this section. In that case, we note the following:

A. $\inf_f \liminf_n \inf_h n^{2/5}\mathbf{E}\{\int |f_{nh} - f|\} \geq A(K)$ where $A(K) \geq 0.86$ is a function of K only, and is minimized for the Epanechnikov kernel $K(x) = \frac{3}{4}(1 - x^2)_+$ (Devroye and Penrod, 1984; see also Devroye and Györfi, 1985). The $n^{-2/5}$ rate is thus a universal lower bound beyond which we cannot go within the class of standard kernel densities.

B. There is no uniform rate of convergence: $\sup_f \inf_h \mathbf{E}\{\int |f_{nh} - f|\} = 2$ (see the exercises).

C. There is no universal rate of convergence for individual densities: for any sequence $a_n \downarrow 0$, there exists a density f such that for all n large enough,

$$\inf_h \mathbf{E}\left\{\int |f_{nh} - f|\right\} > a_n.$$

So, to study rates of convergence, we must assume certain things on the part of the density. For example, certain densities cannot possibly yield an expected error rate of $O(n^{-2/5})$: these include long-tailed densities (densities with $\int \sqrt{f} = \infty$), and not sufficiently smooth densities (e.g., absolutely continuous densities f with absolutely continuous first derivative f', yet $\int |f''| = \infty$). For all these densities, $\liminf_n \inf_h n^{2/5}\mathbf{E}\{\int |f_{nh} - f|\} = \infty$. Let $\mathcal{W}$ be the class of all densities that are absolutely continuous and have an absolutely continuous first derivative f' and have $\int |f''| < \infty$, and assume that $\int \sqrt{f_*} < \infty$ (where $f_*(x) = \sup_{x-1 \leq y \leq x+1} f(y)$). Assume also that $\int x^2 K(x)\, dx < \infty$ and that K is symmetric about the origin. Then

$$\lim_{n\to\infty} n^{2/5} \inf_h \mathbf{E}\left\{\int |f_{nh} - f|\right\} = Q(K,f) < \infty,$$

where $Q(K,f)$ is a rather complicated function of K and f. Luckily, the function $A(K)B(f)$ with $A(K) = \left(\int K^2\right)^{2/5} \left(\int x^2 K(x)\,dx\right)^{1/5}$, $B^5(f) = \frac{1}{2}\left(\int \sqrt{f}\right)^4 \int |f''|$ is a close approximation in the sense that for densities in $\mathcal{W}$,

$$1.028493\ldots \le \frac{Q(K,f)}{A(K)B(f)} \le \frac{5}{\sqrt{2\pi}} = 1.3768102\ldots$$

(Devroye and Györfi, 1985). Interestingly, $B(f)$ is minimal for the triangular density $(1-|x|)_+$ but this density does not minimize $Q(K,f)$ in general (see Devroye and Wand, 1993). Also, $A(K)$ is minimal for the Epanechnikov kernel. Within the class of positive kernels, these results suggest that the Epanechnikov kernel is best. Yet, as shown by Cline (1988, 1990), for the L_2 criterion, there is a better (but not nonnegative) kernel in the sense that for every n and f, the best L_2 error is improved, so that the Epanechnikov kernel is not admissible! In fact, it is rather frustrating to note that for every density in $\mathcal{W}$, and any symmetric kernel L with $\int x^4 |L| < \infty$ and $\int x^2 L = 0$:

$$\lim_{n\to\infty} n^{2/5} \inf_h \mathbf{E}\left\{\int |f_{nh} - f|\right\} = 0.$$

This says that we are unnecessarily limiting ourselves by sticking to nonnegative kernels K. In fact, each kernel creates a best possible rate of convergence, a limiting rate, beyond which we cannot go. This phenomenon of "saturation" is well known in functional approximation. We will provide more details in Chapter 16, where various classes of such kernels will be discussed.

§9.8. Uniform Rate of Convergence

We will need a result that gives nonasymptotic L_1 performance bounds for all densities f in a certain class of densities. We offer one such bound as an example of the sort of calculation that is involved. We note that

$$\sup_{f\in\mathcal{W}} n^{2/5} \inf_h \mathbf{E}\left\{\int |f_{nh} - f|\right\} = \infty$$

so that it is necessary to look at suitable subclasses of $\mathcal{W}$. Consider all densities of compact support, and introduce $s(f)$, the length of the smallest interval of probability one. Consider a further subclass of densities for which f is absolutely continuous with derivative f', which is also absolutely continuous with derivative f'', such that $c(f) \stackrel{\text{def}}{=} \int |f''| < \infty$.

For a constant $C < \infty$, define $\mathcal{W}_C$ as the class of all such densities with $s^2(f)c(f) \leq C$. To some extent, the translation and scale-invariant quantity $s^2(f)c(f)$ measures the difficulty f poses to the kernel estimate.

THEOREM 9.5. *Let the kernel K be bounded, symmetric and supported on $[-1,1]$. Let f_{nh} be the kernel estimate with kernel K and bandwidth h depending upon n only. Then there exists a constant C' depending upon K only such that*

$$\sup_{f\in\mathcal{W}_C} \inf_{h>0} \mathbf{E}\left\{\int |f_{nh} - f|\right\} \leq \frac{C'C^{1/5}}{n^{2/5}}.$$

PROOF. We note that

$$\mathbf{E}\left\{\int |f_{nh} - f|\right\} \leq \int |f * K_h - f| + \mathbf{E}\left\{\int |f_{nh} - f * K_h|\right\}.$$

The terms on the right-hand side will be called the bias and variation terms of the error. By Jensen's inequality,

$$\mathbf{E}\left\{\int |f_{nh} - f|\right\} \geq \int |f * K_h - f|$$

and by Scheffé's identity,

$$\mathbf{E}\left\{\int |f_{nh} - f|\right\} \geq \tfrac{1}{2}\mathbf{E}\left\{\int |f_{nh} - f * K_h|\right\}$$

(see the exercises). Thus, the L_1 error is sandwiched between one half and two times the maximum of the bias and variation terms. As a rule, bias decreases as $h \downarrow 0$ (but this is not always true!), and the variation increases as $h \downarrow 0$ (again, this is not always true). By considering the Taylor series expansion of f about x in $f * K_h(x) = \int f(y)K_h(x-y)\,dy$, stopping at two terms, we see that if $\int xK(x)\,dx = 0$ and $\int x^2K(x)\,dx < \infty$, then

$$|f * K_h(x) - f(x)| \leq h^2 f'' * L_h,$$

where L is the associated kernel of K defined by $L(-x) = L(x)$, and $L(x) = \int_x^\infty (y-x)K(y)\,dy$, $x > 0$. From this, by Young's inequality, and $\int |L| = \int x^2K(x)\,dx/2$, we have

$$\int |f * K_h(x) - f(x)| \leq h^2 \frac{\int |f''| \int x^2 K(x)\,dx}{2} \leq h^2 \frac{c(f)\int x^2 K(x)\,dx}{2}.$$

Next, we turn to the variation term and note that

$$\begin{aligned}\mathbf{E}\{|f_n - f * K_h|\} &\leq \sqrt{\mathbf{E}\{(f_n - f * K_h)^2\}}\\ &\leq \sqrt{\frac{\mathbf{E}\{(K_h(x - X_1))^2\}}{n}}\end{aligned}$$

$$= \frac{\sqrt{M_h * f}}{\sqrt{nh}} \quad \text{(where } M = K^2\text{)}$$

so that, taking integrals and applying the Cauchy–Schwarz inequality and Young's inequality,

$$\begin{aligned} \int \mathbf{E}\{|f_n - f * K_h|\} &\le \frac{\int \sqrt{M_h * f}}{\sqrt{nh}} \\ &\le \frac{\sqrt{s(f) + 2h}\sqrt{\int M_h * f}}{\sqrt{nh}} \\ &\le \frac{\sqrt{s(f) + 2h}\sqrt{\int K^2}}{\sqrt{nh}}. \end{aligned}$$

Take $h = (s(f)/(nc(f)^2))^{1/5}$ and note that for each density in $\mathcal{W}_C$, the L_1 error is bounded by

$$\frac{C' s(f)^{2/5} c(f)^{1/5}}{n^{2/5}},$$

where C' is a constant depending upon K only (and which happens to be minimal for the Epanechnikov kernel if we must select from positive kernels). □

§9.9. Shrinkage, and the Combination of Density Estimates

The kernel estimate is universally consistent, and is our basic workhorse for nonparametric density estimation. However, as the $\Omega(n^{-2/5})$ lower bound shows, its performance is often limited. If $\mathcal{F}$ is a target class of densities, then it is possible to modify the kernel estimate, replacing it with a special purpose estimate, if it is suspected that the true density is in this target class. Three strategies come to mind. In all of these, f_{nh} is the kernel estimate with bandwidth h, and g_n is a special purpose estimate for $\mathcal{F}$ (such as a normal density estimate when $\mathcal{F}$ is normal). We use the notation f_n^* for the combined estimate.

A. One could use the Scheffé method to distinguish between f_n and g_n (Chapter 6). At best, we can deduce that $\mathbf{E}\{\int |f_n^* - f|\}$ comes within a factor of three of the optimal error, but it may often be sufficient.
B. Cao and Devroye (1996) found h and $g_n \in \mathcal{F}$ jointly by minimizing the distance $\int |f_{nh} - g_n|$. Thus, g_n may be considered as the projection of f_{nh} to $\mathcal{F}$. Still, after the minimization, one must decide whether g_n or f_{nh} should be used.
C. Devroye (1989b) suggests a method of combination based upon a thorough knowledge of the properties of f_{nh} and g_n. This method is sketched in this section.

We check the distance between the two estimates themselves, that is, $\int |g_n - f_{nh}|$. The following explicit estimates are assumed known:

$$\sup_{f\in\mathcal{F}} \mathbf{E}\left\{\int |g_n - f|\right\} \le \delta_n;$$

$$\sup_{f\in\mathcal{F}} \mathbf{E}\left\{\int |f_{nh} - f|\right\} \le \alpha_n.$$

Assume also that we know that

$$\sup_{f\in\mathcal{F}} \mathbf{P}\left\{\left|\int |f_{nh} - f| - \mathbf{E}\left\{\int |f_{nh} - f|\right\}\right| \ge \beta_n\right\} \le \gamma_n \to 0$$

for some sequence β_n, and similarly for g_n. Thus, concentration is required.

Define the shrinkage or combined estimate f_n^* as follows:

$$f_n^* = \begin{cases} g_n & \text{if } \int |g_n - f_{nh}| \le 2\beta_n + \alpha_n + \delta_n, \\ f_{nh} & \text{otherwise.} \end{cases}$$

Theorem 6 below shows that on $\mathcal{F}$, if $\gamma_n = o(\delta_n)$, we match the performance of g_n. However, outside $\mathcal{F}$, disasters are avoided, and we obtain universal consistency, often with guaranteed rates of convergence. Examples are suggested in the exercises. Needless to say, there is no particular reason why we should take f_{nh} as the kernel estimate. Any other general purpose estimate such as the histogram estimate, would have been equally useful.

THEOREM 9.6. *For the shrinkage estimate f_n^*, if $f \in \mathcal{F}$:*

$$\mathbf{E}\left\{\int |f_n^* - f|\right\} \le \delta_n + 4\gamma_n.$$

And for any f:

$$\mathbf{E}\left\{\int |f_n^* - f|\right\} \le 2\mathbf{E}\left\{\int |f_{nh} - f|\right\} + 2\beta_n + \alpha_n + \delta_n.$$

PROOF. Assume that $f \in \mathcal{F}$. Then

$$\int |g_n - f_{nh}| \le \int |g_n - f| + \int |f_{nh} - f| \le \delta_n + \beta_n + \alpha_n + \beta_n$$

with probability at least $1 - 2\gamma_n$. Thus, g_n is picked with high probability. Therefore,

$$\begin{aligned} \mathbf{E}\left\{\int |f_n^* - f|\right\} &\le \mathbf{E}\left\{\int |g_n - f|\right\} \\ &\quad + \mathbf{E}\left\{\int |f_{nh} - f| \mathbf{1}_{[f_{nh} \text{ picked}]}\right\} \\ &\le \delta_n + 4\gamma_n. \end{aligned}$$

Outside $\mathcal{F}$, the error of the combined estimate is bounded as follows:

$$\begin{aligned}
&\mathbf{E}\left\{\int |f_n^* - f|\right\} \\
&\quad \le \mathbf{E}\left\{\int |f_{nh} - f|\right\} + \mathbf{E}\left\{\int |g_n - f| \mathbf{1}_{[g_n \text{ picked}]}\right\} \\
&\quad \le \mathbf{E}\left\{\int |f_{nh} - f|\right\} + \mathbf{E}\left\{\int |f_{nh} - f| \mathbf{1}_{[g_n \text{ picked}]}\right\} \\
&\qquad + \mathbf{E}\left\{\int |f_{nh} - g_n| \mathbf{1}_{[g_n \text{ picked}]}\right\} \\
&\quad \le 2\mathbf{E}\left\{\int |f_{nh} - f|\right\} + 2\beta_n + \alpha_n + \delta_n. \ \square
\end{aligned}$$

§9.10. Bibliographic Remarks

The variable kernel estimates were first investigated by Breiman, Meisel and Purcell (1977). Their L_1 convergence was studied by Devroye (1985). The necessity of the conditions in Theorem 2 was proved by Devroye (1983) (see also the exercises). Theorems 3 and 5 are due to Devroye and Györfi (1985), and so are the necessary conditions for obtaining the rate $O(n^{-2/5})$. The relative stability of the kernel estimate was first studied by Devroye (1987, 1988a, 1991). The concentration inequality in Theorem 4 shows that $\int |f_n - f| - \mathbf{E}\int |f_n - f|$ is of the order of $O(1/\sqrt{n})$. In fact, $\sqrt{n}(\int |f_n - f| - \mathbf{E}\int |f_n - f|)$ tends to a normal limit (Csörgő and Horváth, 1988; Beirlant and Mason, 1995). A detailed study of the L_1 behavior of the kernel estimate in $\mathbf{R}^d$ is provided by Holmström and Klemelä (1992). Bandwidth selection methods are surveyed by Berlinet and Devroye (1994), Cao, Cuevas and González-Manteiga (1994), and Devroye (1997). In particular, the maximum likelihood cross-validation method mentioned in the text was studied in detail by Broniatowski, Deheuvels and Devroye (1989). The combination of nonparametric estimates with nonaparametric and/or parametric estimates is done in the text along the lines of Devroye (1989b) and Cao and Devroye (1996).

§9.11. Exercises

Exercise 9.1 Prove the following generalization of Theorem 1: let f be a density on $\mathbf{R}$, let K be a function on $\mathbf{R}$ of integral one, and define $L_{h_1,\ldots,h_d}(x_1,\ldots,x_d) = K_{h_1}(x_1) \times \cdots K_{h_d}(x_d)$, $x_i \in \mathbf{R}, 1 \le i \le d$, then

$$\lim_{\max_i h_i \downarrow 0} \int |f * L_{h_1,\ldots,h_d} - f| = 0.$$

NOTE: The relative rates of convergence of the h_i's to zero may be very different. There is a proof based on coupling.

Exercise 9.2 Prove Theorem 1 when $\int K = 0$.

Exercise 9.3 Consider a kernel estimate for which $\mathbf{E}\{\int |f_{nh} - f|\} \to 0$ for a particular density f. Show that $nh^d \to \infty$.

Exercise 9.4 Let $d = 1$. Consider a kernel estimate with a kernel K for which for some integer $s > 0$, $\int x^s K(x)\,dx \neq 0$. Show that if $\mathbf{E}\{\int |f_n - f|\} \to 0$, then $h \to 0$. NOTE: For such kernels on the real line, this result together with the previous exercise and Theorem 2 show that the following conditions are equivalent:

A. $\mathbf{E}\{\int |f_n - f|\} \to 0$ for one density f;
B. $\mathbf{E}\{\int |f_n - f|\} \to 0$ for all densities f;
C. $h \to 0$ and $nh^d \to \infty$.

Exercise 9.5 Prove Theorem 3. NOTE: This is not as easy as it sounds, and should only be attempted by the brave.

Exercise 9.6 Consider the double kernel estimate on $\mathbf{R}$ with two kernels having characteristic functions that do not coincide on an open interval about the origin. If H is such that $\int |f_{n,H} - g_{n,H}| \sim \inf_h \int |f_{n,h} - g_{n,h}|$, then show that $H \to 0$ in probability. HINT: Get a contradiction. The following inequality may be useful: for any functions f, g with characteristic function ϕ, ψ, we have $\int |f - g| \geq \sup_t |\phi(t) - \psi(t)|$.

Exercise 9.7 There is no uniform rate of convergence for the kernel estimate on $\mathbf{R}^1$: show that $\sup_f \inf_h \mathbf{E}\{\int |f_{nh} - f|\} = 2$.

Exercise 9.8 Show that there is no universal rate of convergence for individual densities: for any sequence $a_n \downarrow 0$, there exists a density f such that $\inf_h \mathbf{E}\{\int |f_{nh} - f|\} > a_n$ for all n large enough. Show, in particular, that f may be chosen from each of the following classes:

A. The class of all bounded monotone densities on $[0, \infty)$ with infinitely many continuous derivatives on $(0, \infty)$.
B. The class of densities on $[0, 1]$ that are bounded by 2.
C. The class of densities of the form $g * \nu$, where g is the standard normal density, and ν is an arbitrary probability measure.

Exercise 9.9 Show that in the L_1 metric, the following classes of densities are dense in the class of all densities:

A. The class of all bounded densities with compact support.
B. The class of all Lipschitz densities with compact support.

C. The class of all densities on $\mathbf{R}$ with compact support and k uniformly continuous derivatives, where $k \geq 1$ is an arbitrary integer.

Exercise 9.10 Let K be the product of d univariate kernels. Show that each marginal of the kernel density estimate is a kernel density estimate with product kernel. If K is an arbitrary kernel, then show that the marginals of the kernel density estimate are still bona fide kernel density estimates. In case K is the uniform density on the unit ball of $\mathbf{R}^d$, show that the kernels of the univariate marginal density estimates are symmetric beta densities.

Exercise 9.11 Show that for every density in $\mathcal{W}$, and any symmetric kernel L with $\int x^4|L| < \infty$ and $\int x^2 L = 0$:

$$\lim_{n\to\infty} n^{2/5} \inf_h \mathbf{E}\left\{\int |f_{nh} - f|\right\} = 0.$$

HINT: Show that the bias is $o(h^2)$. The variation term is $O(1/\sqrt{nh})$.

Exercise 9.12 Assume that K is the uniform kernel on $[-1,1]$. Show that the maximum likelihood bandwidth H must be larger than the maximal spacing between consecutive data points. If f is such that $f(x)e^x \to \infty$ as $x \to \infty$, show that there exist constants $a, p > 0$ such that $\mathbf{P}\{H \geq a\} \geq p > 0$ for all n large enough. Finally, show that $\mathbf{E}\{\int |f_{n,H} - f|\}$ does not tend to zero.

Exercise 9.13 Consider $J_{n,h} = \mathbf{E}\{\int (f_{n,h} - f)^2\}$ where $f_{n,h}$ is the standard kernel estimate on $\mathbf{R}$ with bandwidth h and kernel K satisfying $\int x^2 K = 1$, $\int K^2 = \alpha < \infty$, and where f is a bounded density with two continuous derivatives and $\int f''^2 = c < \infty$. By splitting into bias and variance, show that $\inf_h J_{n,h} \sim 5\alpha^{4/5}c^{1/5}/4n^{4/5}$ and prove that this infimum is asymptotically attained for $h = bn^{-1/5}$, where b is a function of α and c only. Determine b (Rosenblatt, 1971; Nadaraya, 1974).

Exercise 9.14 Consider the Cauchy density $f(x) = 1/(\pi(1+x^2))$ on the real line. Determine the order of decrease to zero of $\inf_h \mathbf{E}\{\int |f_{n,h} - f|\}$. Determine the rate of decrease to zero of $\mathbf{E}\{\int |f_{n,h} - f|\}$ when $h = bn^{-1/5}$ for constant $b > 0$. Using the result of the previous exercise, conclude that for the Cauchy density, the L_2-optimal bandwidth leads to an unacceptable L_1 error.

Exercise 9.15 Consider the standard kernel estimate in $\mathbf{R}^d$ with kernel K symmetric in all components, $K \geq 0$ and $\int \|x\|^2 K(x)\,dx < \infty$. Using the multivariate Taylor series expansion of f to bound the bias, find sufficient conditions on f (that somehow generalize the sufficient conditions for f noted in the text for $d = 1$) for the expected L_1 error to be $O(1/n^{2/(4+d)})$.

Exercise 9.16 Show that for i.i.d. random variables $X_1, \ldots, X_n$, any nonnegative convex function ψ, and any probability vector $(p_1, \ldots, p_n)$:

$$\mathbf{E}\left\{\psi\left(\sum_{i=1}^{n} p_i X_i\right)\right\} \geq \mathbf{E}\left\{\psi\left(\frac{1}{n}\sum_{i=1}^{n} X_i\right)\right\}.$$

Exercise 9.17 From the previous exercise, conclude that for the standard kernel estimate with fixed bandwidth $h > 0$, $\mathbf{E}\{\int |f_{nh} - f|\}$ is nonincreasing in n.

Exercise 9.18 Let K be the uniform density on $[-1, 1]$. Show that there exists a density f such that the bias of the kernel estimate, $\int |f - f * K_h|$ is not nonincreasing as $h \to 0$. However, if K is the standard normal density, and f is any normal density, show that the bias is indeed monotone.

Exercise 9.19 Let f_{nh} be the standard kernel estimate with kernel K. Show that

$$\mathbf{E}\left\{\int |f_{nh} - f|\right\} \geq \tfrac{1}{2}\mathbf{E}\left\{\int |f_{nh} - f * K_h|\right\}.$$

Exercise 9.20 Let f_{nh} and g_{nh} be two standard kernel estimates based on independent samples of size n, but using the same bandwidth h and kernel K. Show that

$$\mathbf{E}\left\{\int |f_{nh} - f|\right\} \geq \tfrac{1}{2}\mathbf{E}\left\{\int |f_{nh} - g_{nh}|\right\}.$$

Exercise 9.21 The objective of this exercise is to show that a bias estimate in the proof of Theorem 5 is tight. Assume that $f \in \mathcal{W}_C$. In the notation of that proof, show the following steps:

A. If L is the associated kernel defined in the proof of Theorem 5, show the following, by using Taylor series expansions:

$$|f * K_h(x) - f(x)| < h^2 f'' * L_h.$$

B. Let x be a Lebesgue point for f'' (note that formally, Theorem 5.3 remains valid for all integrable functions f). Using Theorem 5.3, show that $f'' * L_h(x) \to f''(x) \int L$ as $h \to 0$.

C. Conclude that

$$\lim_{h\to 0} \frac{|f * K_h(x) - f(x)|}{h^2} = \lim_{h\to 0} |f'' * L_h| = |f''(x)| \int |L|.$$

D. Apply Fatou's lemma to show that

$$\liminf_{h\to 0} \frac{\int |f * K_h(x) - f(x)|}{h^2} \geq \int |f''| \int |L|.$$

Exercise 9.22 For each $a \in (0,2)$, find a monotonically decreasing density f on $[0,1]$ for which the bias $\int |f - f * K_h|$ is asymptotic to h^a as $h \to 0$, where K is the uniform density on $[-1,1]$. Then for any function $\psi(h)$ that decreases monotonically to 0 with h, find a monotonically decreasing density f on $[0,1]$ for which the bias is $\geq \psi(h)$ for all h large enough. (This will convince you that any slow rate for the bias is possible just by considering densities with one infinite peak.)

Exercise 9.23 Let K be the uniform kernel on $[-1,1]$. Find a density f for which $\int \sqrt{f} < \infty$, all derivatives are continuous, $\int |f''| < \infty$, yet $n^{2/5} \inf_h \mathbf{E}\{\int |f_{nh} - f|\} \to \infty$. HINT: The variation term may be much larger than $1/\sqrt{nh}$ if f has many well-separated bumps.

Exercise 9.24 Consider the class of densities f on the real line with support $s(f)$ and satisfying a Lipschitz condition: $\alpha(f) \stackrel{\text{def}}{=} \sup_{x\neq y} |f(x) - f(y)|/|x-y| < \infty$. Show that the bias of the L_1 error is not greater than $2\alpha(f)h \int |x||K(x)|\,dx$. Let $\mathcal{L}_C = \{f : \alpha(f)s^2(f) \leq C^3\}$ for a given constant C. Show that if $\int x^2|K| < \infty$, then

$$\sup_{f\in A_C} \mathbf{E}\left\{\int |f_{nh} - f|\right\} \leq \frac{CC'}{n^{1/3}},$$

where C' is a finite constant depending upon K only.

Exercise 9.25 Repeat the previous exercise for bounded densities on hyperrectangles of $\mathbf{R}^d$ that are monotone in each coordinate separately. Determine the exponent of n in the uniform rate of convergence for the class of densities for which the volume of the support hyperrectangle times c, the bound on f, is not more than a given constant C.

Exercise 9.26 Let f be a monotone density on $[a,b]$, bounded by c. Let K be the uniform density on $[-1,1]$. Show that the rate of convergence of the expected L_1 error of the standard kernel estimate is $O(n^{-1/3})$. Show, furthermore, that this rate is uniform over all such densities with $(b-a)c \leq C$, where $C > 1$ is a fixed constant.

Exercise 9.27 Let f be a unimodal density on the real line with $m(f) \stackrel{\text{def}}{=} \sup_x f(x) < \infty$, and variance $\sigma^2(f) < \infty$. Let K be the uniform density on $[-1,1]$. Show that the rate of convergence of the expected L_1 error of the standard kernel estimate is $O(n^{-1/3})$. Show furthermore, that this rate is uniform over all such densities with $m(f)(1+\sigma^2(f)) \leq$

C, where $C > 1$ is a fixed constant. HINT: First prove the inequality $\int \sqrt{f} \le \sqrt{\pi(1+\sigma^2(f))}$ by multiplying and dividing $\sqrt{f}$ by $\sqrt{1+x^2}$, and applying the Cauchy–Schwarz inequality.

Exercise 9.28 Establish a uniform upper bound for $\inf_h \mathbf{E}\{\int |f_{nh} - f|\}$ over the following classes:

A. The class of all log-concave densities on the real line (i.e., $\log f$ is concave).
B. The class of all concave densities on a compact interval.

Exercise 9.29 (SHRINKAGE ESTIMATE EXAMPLE). Let $\mathcal{F}$ be the subclass of all uniform densities on $[a,b]$, for all possible values of $[a,b]$. The density estimate g_n is the uniform density on $[\min_i X_i, \max_i X_i]$, where $X_1, \ldots, X_n$ is an i.i.d. sample of size n drawn from the unknown density f.

A. Verify that $\sup_{f\in\mathcal{F}} \mathbf{E}\{\int |g_n - f|\} \le C/n$ for a universal constant C.
B. Let f_{nh} be the kernel estimate with bandwidth $h = S/n^{1/3}$, where S is the spread of the X_i's, i.e., the difference between the 75% and 25% quantiles of the data. Show that f_{nh} is consistent for all densities.
C. Let f_n^* be the shrinkage estimate defined in the text. Determine α_n, β_n and γ_n for use in Theorem 6 such that for the shrinkage estimate, when $f \in \mathcal{F}$, $\mathbf{E}\{\int |f_n^* - f|\} \le (C+o(1))/n$, and that when $f \notin \mathcal{F}$, $\mathbf{E}\{\int |f_n^* - f|\} \le \mathbf{E}\{\int |f_{nh} - f|\} + O(1/n^{1/3})$.

§9.12. References

H. Akaike, "An approximation to the density function," *Annals of the Institute of Statistical Mathematics*, vol. 6, pp. 127–132, 1954.

J. Beirlant and D. M. Mason, "On the asymptotic normality of L_p-norms of empirical functionals," *Mathematical Methods of Statistics*, vol. 4, pp. 1–19, 1995.

A. Berlinet and L. Devroye, "A comparison of kernel density estimates," *Publications de l'Institut de Statistique de l'Université de Paris*, vol. 38, pp. 3–59, 1994.

L. Breiman, W. Meisel, and E. Purcell, "Variable kernel estimates of multivariate densities," *Technometrics*, vol. 19, pp. 135–144, 1977.

M. Broniatowski, P. Deheuvels, and L. Devroye, "On the relationship between stability of extreme order statistics and convergence of the maximum likelihood kernel density estimate," *Annals of Statistics*, vol. 17, pp. 1070–1086, 1989.

R. Cao, A. Cuevas, and W. González–Manteiga, "A comparative study of several smoothing methods in density estimation," *Computational Statistics and Data Analysis*, vol. 17, pp. 153–176, 1994.

R. Cao and L. Devroye, "The consistency of a smoothed minimum distance estimate," *Scandinavian Journal of Statistics*, vol. 23, pp. 405–418, 1996.

D. B. H. Cline, "Admissible kernel estimators of a multivariate density," *Annals of Statistics*, vol. 16, pp. 1421–1427, 1988.

D. B. H. Cline, "Optimal kernel estimation of densities," *Annals of the Institute of Statistical Mathematics*, vol. 42, pp. 287–303, 1990.

M. Csörgő and L. Horváth, "Central limit theorems for L_p-norms of density estimators," *Probability Theory and Related Fields*, vol. 80, pp. 269–291, 1988.

P. Deheuvels, "Estimation nonparamétrique de la densité par histogrammes generalisés," *Publications de l'Institut de Statistique de l'Université de Paris*, vol. 22, pp. 1–23, 1977.

L. Devroye, "The equivalence of weak, strong and complete convergence in L_1 for kernel density estimates," *Annals of Statistics*, vol. 11, pp. 896–904, 1983.

L. Devroye, "A note on the L_1 consistency of variable kernel estimates," *Annals of Statistics*, vol. 13, pp. 1041–1049, 1985.

L. Devroye, "The kernel estimate is relatively stable," *Probability Theory and Related Fields*, vol. 77, pp. 521–536, 1988a.

L. Devroye, "Asymptotic performance bounds for the kernel estimate," *Annals of Statistics*, vol. 16, pp. 1162–1179, 1988b.

L. Devroye, "The double kernel method in density estimation," *Annales de l'Institut Henri Poincaré*, vol. 25, pp. 533–580, 1989a.

L. Devroye, "Nonparametric density estimates with improved performance on given sets of densities," *Statistics (Mathematische Operationsforschung und Statistik)*, vol. 20, pp. 357–376, 1989b.

L. Devroye, "Exponential inequalities in nonparametric estimation," in: *Nonparametric Functional Estimation and Related Topics* (edited by G. Roussas), pp. 31–44, NATO ASI Series, Kluwer Academic, Dordrecht, 1991.

L. Devroye, "Universal smoothing factor selection in density estimation: Theory and practice (with discussion)," *Test*, vol. 6, pp. 223–320, 1997.

L. Devroye, *A Course in Density Estimation*, Birkhäuser-Verlag, Boston, 1987.

L. Devroye and L. Györfi, *Nonparametric Density Estimation: The L_1 View*, John Wiley, New York, 1985.

L. Devroye and C. S. Penrod, "Distribution-free lower bounds in density estimation," *Annals of Statistics*, vol. 12, pp. 1250–1262, 1984.

L. Devroye and M. P. Wand, "On the effect of density shape on the performance of its kernel estimate," *Statistics*, vol. 24, pp. 215–233, 1993.

V. A. Epanechnikov, "Nonparametric estimation of a multivariate probability density," *Theory of Probability and its Applications*, vol. 14, pp. 153–158, 1969.

P. Hall and M. P. Wand, "Minimizing L_1 distance in nonparametric density estimation," *Journal of Multivariate Analysis*, vol. 26, pp. 59–88, 1988.

L. Holmström and J. Klemelä, "Asymptotic bounds for the expected L_1 error of a multivariate kernel density estimator," *Journal of Multivariate Analysis*, vol. 40, pp. 245–255, 1992.

E. A. Nadaraya, "On the integral mean square error of some nonparametric estimates for the density function," *Theory of Probability and its Applications*, vol. 19, pp. 133–141, 1974.

E. Parzen, "On the estimation of a probability density function and the mode," *Annals of Mathematical Statistics*, vol. 33, pp. 1065–1076, 1962.

M. Rosenblatt, "Remarks on some nonparametric estimates of a density function," *Annals of Mathematical Statistics*, vol. 27, pp. 832–837, 1956.

M. Rosenblatt, "Curve estimates," *Annals of Mathematical Statistics*, vol. 42, pp. 1815–1842, 1971.

C. J. Stone, "An asymptotically optimal window selection rule for kernel density estimates," *Annals of Statistics*, vol. 12, pp. 1285–1297, 1984.

G. S. Watson and M. R. Leadbetter, "On the estimation of the probability density, I," *Annals of Mathematical Statistics*, vol. 34, pp. 480–491, 1963.

M. Wegkamp, "Quasi universal bandwidth selection for kernel density estimators," *Canadian Journal of Statistics*, 2000. To appear.

Chapter 10
Additive Estimates and Data Splitting

§10.1. Data Splitting

Assume that we are given a class of density estimates parametrized by $\theta \in \Theta$, such that $f_{n,\theta}$ denotes the density estimate with parameter θ. Our goal is to construct a density estimate f_n whose L_1 error is (almost) as small as that of the best estimate among the $f_{n,\theta}$, $\theta \in \Theta$. Applying the minimum distance estimate of Chapter 5 directly to this class is often problematic because of the dependence of each estimate in the class and the empirical measure μ_n. Consider, as a basic example, the class of kernel estimates

$$f_{n,h}(x) = \frac{1}{nh^d} \sum_{i=1}^{n} K\left(\frac{x - X_i}{h}\right)$$

parametrized by the smoothing factor $h \in (0, \infty)$, where K is a fixed nonnegative function with $\int K = 1$. It is easy to see that if

$$\mathcal{A} = \{\{f_{n,h} > f_{n,h'}\} : h, h' \in (0, \infty), h \neq h'\},$$

then

$$\sup_{A \in \mathcal{A}} \left| \int_A f_{n,h} - \mu_n(A) \right|$$

converges to zero as $h \to 0$, so the minimum distance method selects a degenarate estimate.

We remedy this problem by introducing artificial independence between the estimates and the empirical measure. This may be achieved by holding out m samples from the design of the density estimates, and using the empirical measure based on the held-out samples to construct the minimum distance estimate. More precisely, let $m < n$, and define $\mathcal{A}_\Theta$ as the *Yatracos class* of subsets of $\mathbf{R}^d$ (corresponding to the family of density estimates $f_{n,\theta}$, $\theta \in \Theta$) as the class of all sets of the form

$$A_{\theta_1,\theta_2} = \{x : f_{n-m,\theta_1}(x) > f_{n-m,\theta_2}(x)\}, \quad \theta_1, \theta_2 \in \Theta.$$

We select a parameter θ_n from Θ by minimizing the distance

$$\Delta_\theta = \sup_{A \in \mathcal{A}_\Theta} \left| \int_A f_{n-m,\theta} - \mu_m(A) \right|.$$

over all $\theta \in \Theta$, where μ_m denotes the empirical measure defined by the subsample $X_{n-m+1}, \ldots, X_n$. If the minimum does not exist, we select θ_n such that $\Delta_{\theta_n} < \inf_{\theta^* \in \Theta} \Delta_{\theta^*} + 1/n$. Define $f_n = f_{n-m,\theta_n}$. Using Theorem 6.4 conditionally, we immediately obtain:

THEOREM 10.1. *If $\int f_{n-m,\theta} = 1$ for all $\theta \in \Theta$, then for the minimum distance estimate f_n as defined above, we have*

$$\int |f_n - f| \le 3 \inf_{\theta \in \Theta} \int |f_{n-m,\theta} - f| + 4\Delta + \frac{3}{n},$$

where

$$\Delta = \sup_{A \in \mathcal{A}_\Theta} \left| \int_A f - \mu_m(A) \right|.$$

Note that if the estimates do not satisfy the condition $\int f_{n-m,\theta} = 1$, then Theorem 1 still remains valid, but with the factors of "3" replaced by "5" (Exercise 6.2). In order to use Theorem 1, first we have to make sure that $\inf_{\theta \in \Theta} \int |f_{n-m,\theta} - f|$ is not much larger than $\inf_{\theta \in \Theta} \int |f_{n,\theta} - f|$, that is, holding out m samples does not hurt. In the next section we will see that for many important families of estimates, this is indeed the case. The second part of the analysis is then purely combinatorial, as upper bounds for the value of Δ may be obtained by bounding the shatter coefficients of the class $\mathcal{A}_\Theta$.

§10.2. Additive Estimates

Many classical nonparametric density estimates may be written in the form

$$g_n(x) = \frac{1}{n} \sum_{i=1}^{n} K(x, X_i),$$

where $K : \mathbf{R}^d \times \mathbf{R}^d \to \mathbf{R}$ is a measurable function. Such estimates are called *additive*. We say that the additive estimate g_n is *regular* if for each x, $\mathbf{E}\{|K(x, X)|\} < \infty$. Examples of additive estimates include the kernel (Chapters 11, 16 and 17), histogram (this chapter), series, and wavelet estimates (Chapter 13). The key result for families of such estimates is Theorem 2.

THEOREM 10.2. *Let Θ be a class of parameters, and assume that each density estimate $f_{n,\theta}(x) = (1/n) \sum_{i=1}^{n} K_\theta(x, X_i)$ is additive and regular. Denote $J_{n,\theta} = \int |f_{n,\theta} - f|$. If $m > 0$ is a positive integer such that $2m \le n$, then*

$$\frac{\inf_{\theta \in \Theta} \mathbf{E}\{J_{n-m,\theta}\}}{\inf_{\theta \in \Theta} \mathbf{E}\{J_{n,\theta}}\} \le 1 + \frac{2m}{n-m} + 8\sqrt{\frac{m}{n}}.$$

This means that by decreasing the sample size to $n - m$, the performance of the best estimate in the class cannot deteriorate by more than a constant factor. If m is small relative to n, the loss in the L_1 error is negligible. The proof uses some simple results about sums of independent random variables developed in Lemmas 1, 2, and 3.

LEMMA 10.1. *Let X and Y be independent random variables, and let $\mathbf{E}\{Y\} = 0$. Then $\mathbf{E}\{|X+Y|\} \geq \mathbf{E}\{|X|\}$.*

PROOF. We write $X = \mathbf{E}\{X+Y|X\}$, and use Jensen's inequality:

$$\mathbf{E}\{|X|\} = \mathbf{E}\{|\mathbf{E}\{X+Y|X\}|\} \leq \mathbf{E}\{|X+Y|\}. \quad \square$$

LEMMA 10.2 (KHINCHINE'S INEQUALITY). *Let $a_1, \ldots, a_n$ be real numbers, and let $\sigma_1, \ldots, \sigma_n$ be i.i.d. sign variables with $\mathbf{P}\{\sigma_1 = 1\} = \mathbf{P}\{\sigma_1 = -1\} = 1/2$. Then*

$$\mathbf{E}\left\{\left|\sum_{i=1}^n a_i\sigma_i\right|\right\} \geq \frac{1}{\sqrt{2}}\sqrt{\sum_{i=1}^n a_i^2}.$$

Compare this with the closely related upper bound

$$\mathbf{E}\left\{\left|\sum_{i=1}^n a_i\sigma_i\right|\right\} \leq \sqrt{\mathbf{E}\left\{\left(\sum_{i=1}^n a_i\sigma_i\right)^2\right\}} = \sqrt{\sum_{i=1}^n a_i^2}.$$

PROOF. The proof of the inequality may be found in Szarek (1976). Here we give a short proof with a suboptimal constant ($1/\sqrt{3}$ instead of $1/\sqrt{2}$). First note that for any random variable X with finite fourth moment,

$$\mathbf{E}\{|X|\} \geq \frac{\left(\mathbf{E}\{X^2\}\right)^{3/2}}{\left(\mathbf{E}\{X^4\}\right)^{1/2}}.$$

Indeed, for any $a > 0$, the function $1/x + ax^2$ is minimal on $(0, \infty)$ when $x^3 = 1/(2a)$. Thus,

$$\frac{x + ax^4}{x^2} \geq (2a)^{1/3} + \frac{a}{(2a)^{2/3}} = 3/2(2a)^{1/3}.$$

Replace x by $|X|$ and take expectations:

$$\mathbf{E}\{|X|\} \geq 3/2(2a)^{1/3}\mathbf{E}\{X^2\} - a\mathbf{E}\{X^4\}.$$

The lower bound, considered as a function of a, is maximized if we take $a = 1/2\left(\mathbf{E}\{X^2\}/\mathbf{E}\{X^4\}\right)^{3/2}$. Resubstitution yields the claim. Applying the above inequality for $X = \sum_{i=1}^n a_i\sigma_i$ gives

$$\mathbf{E}\left\{\left|\sum_{i=1}^n a_i\sigma_i\right|\right\} \geq \frac{\left(\sum_{i=1}^n a_i^2\right)^{3/2}}{\sqrt{\sum_{i=1}^n a_i^4 + 3\sum_{i\neq j} a_i^2 a_j^2}} \geq \frac{1}{\sqrt{3}}\sqrt{\sum_{i=1}^n a_i^2},$$

where we used $\sum_{i=1}^n a_i^4 + 3\sum_{i\neq j} a_i^2 a_j^2 \leq 3(\sum_{i=1}^n a_i^2)^2$. □

Finally, we generalize the above inequality for general sums of independent random variables:

LEMMA 10.3. *Let $Y_1, \ldots, Y_n$ be i.i.d. zero mean random variables. Then*

$$\mathbf{E}\left\{\left|\sum_{i=1}^n Y_i\right|\right\} \geq \sqrt{\frac{n}{8}}\mathbf{E}\{|Y_1|\}.$$

PROOF. The proof uses a symmetrization trick similar to the one used in Theorem 3.1. We introduce the i.i.d. random variables $Y_1', \ldots, Y_n'$, distributed as the Y_i's, and independent of them. Let $\sigma_1, \ldots, \sigma_n$ be i.i.d. sign variables with $\mathbf{P}\{\sigma_1 = -1\} = \mathbf{P}\{\sigma_1 = 1\} = 1/2$, independent of $Y_1, Y_1', \ldots, Y_n, Y_n'$. Then

$$\begin{aligned}
\mathbf{E}\left\{\left|\sum_{i=1}^n Y_i\right|\right\} &= 1/2\left(\mathbf{E}\left\{\left|\sum_{i=1}^n Y_i\right|\right\} + \mathbf{E}\left\{\left|\sum_{i=1}^n Y_i'\right|\right\}\right) \\
&\geq 1/2\,\mathbf{E}\left\{\left|\sum_{i=1}^n (Y_i - Y_i')\right|\right\} \\
&= 1/2\,\mathbf{E}\left\{\left|\sum_{i=1}^n \sigma_i(Y_i - Y_i')\right|\right\} \\
&= 1/2\,\mathbf{E}\left\{\mathbf{E}\left\{\left|\sum_{i=1}^n \sigma_i(Y_i - Y_i')\right| \,\middle|\, Y_1, Y_1', \ldots, Y_n, Y_n'\right\}\right\} \\
&\geq \frac{1}{2\sqrt{2}}\mathbf{E}\left\{\sqrt{\sum_{i=1}^n (Y_i - Y_i')^2}\right\} \quad \text{(by Lemma 2)} \\
&\geq \frac{\sqrt{n}}{2\sqrt{2}}\mathbf{E}\{|Y_i - Y_i'|\} \\
&\geq \frac{\sqrt{n}}{2\sqrt{2}}\mathbf{E}\{|Y_i|\} \quad \text{(by Lemma 1)}. \;\square
\end{aligned}$$

LEMMA 10.4. *For any density estimate g_n,*

$$\mathbf{E}\left\{\int |f - g_n|\right\} \geq 1/2\,\mathbf{E}\left\{\int |g_n - \mathbf{E}g_n|\right\}.$$

PROOF. Sum the two inequalities

$$\mathbf{E}\left\{\int |f - g_n|\right\} \geq \int |f - \mathbf{E}g_n| \quad \text{(Jensen's inequality)},$$

and

$$\mathbf{E}\left\{\int |f - g_n|\right\} \geq \mathbf{E}\left\{\int |g_n - \mathbf{E}g_n|\right\} - \int |f - \mathbf{E}g_n| \quad \text{(triangle inequality).} \square$$

PROOF OF THEOREM 2. Note the following:

$$\begin{aligned}
\inf_{\theta\in\Theta} \mathbf{E}\{J_{n-m,\theta}\} &\leq \inf_{\theta\in\Theta} \mathbf{E}\{J_{n,\theta}\} \times \sup_{\theta\in\Theta} \left(\frac{\mathbf{E}\{J_{n-m,\theta}\}}{\mathbf{E}\{J_{n,\theta}}\} \right) \\
&= \inf_{\theta\in\Theta} \mathbf{E}\{J_{n,\theta}\} \times \left(1 + \sup_{\theta\in\Theta} \frac{\mathbf{E}\{J_{n-m,\theta} - \mathbf{E}\{J_{n,\theta}\}\}}{\mathbf{E}\{J_{n,\theta}\}} \right).
\end{aligned}$$

The supremum is rewritten as follows:

$$\begin{aligned}
\sup_{\theta\in\Theta} \frac{\mathbf{E}\{J_{n-m,\theta} - \mathbf{E}\{J_{n,\theta}\}\}}{\mathbf{E}\{J_{n,\theta}\}} &\leq \sup_{\theta\in\Theta} \frac{\mathbf{E}\{\int |f_{n-m,\theta} - f_{n,\theta}|\,dx\}}{\mathbf{E}\{J_{n,\theta}\}} \\
&\leq 2 \sup_{\theta\in\Theta} \frac{\mathbf{E}\{\int |f_{n-m,\theta} - f_{n,\theta}|\,dx\}}{\mathbf{E}\{\int |f_{n,\theta} - \mathbf{E}f_{n,\theta}|\,dx\}},
\end{aligned}$$

where we used Lemma 4. Fix x and θ for now. Introduce

$$Y_i = K_\theta(x, X_i) - \mathbf{E}\{K_\theta(x, X)\},$$

and denote the partial sums of Y_i's by $S_j = Y_1 + \cdots + Y_j$. By assumption, for fixed x and θ, the first absolute moment of Y_1 is finite. Then observe the following:

$$n|f_{n-m,\theta} - f_{n,\theta}| = \left| \frac{m}{n-m}(Y_1 + \cdots + Y_{n-m}) - (Y_{n-m+1} + \cdots + Y_n) \right|$$

so that

$$\mathbf{E}\left\{n|f_{n-m,\theta} - f_{n,\theta}|\right\} \leq \frac{m}{n-m}\mathbf{E}\{|S_{n-m}|\} + \mathbf{E}\{|S_m|\}.$$

Also, $n|f_{n,\theta} - \mathbf{E}f_{n,\theta}| = |S_n|$, which implies $\mathbf{E}\left\{n|f_{n,\theta} - \mathbf{E}f_{n,\theta}|\right\} = \mathbf{E}\{|S_n|\}$. Still holding x and θ fixed, we bound the following ratio:

$$\begin{aligned}
\frac{\mathbf{E}\{|f_{n-m,\theta} - f_{n,\theta}|\}}{\mathbf{E}\{|f_{n,\theta} - \mathbf{E}f_{n,\theta}|\}} &\leq \frac{(m/(n-m)\mathbf{E}\{|S_{n-m}|\} + \mathbf{E}\{|S_m|\}}{\mathbf{E}\{|S_n|\}} \\
&\leq \frac{m}{n-m} + \frac{\mathbf{E}\{|S_m|\}}{\mathbf{E}\{|S_n|\}} \\
&\qquad \text{(because } \mathbf{E}\{|S_n|\} \geq \mathbf{E}\{|S_{n-m}|\}\text{)} \\
&\leq \frac{m}{n-m} + \frac{\mathbf{E}\{|S_m|\}}{\sqrt{\lfloor n/m \rfloor / 8}\mathbf{E}\{|S_m|\}} \quad \text{(by Lemmas 1 and 3)} \\
&\leq \frac{m}{n-m} + 4\sqrt{\frac{m}{n}} \quad \text{(if } 2m \leq n\text{)}.
\end{aligned}$$

This implies that for any fixed θ:

$$\mathbf{E}\left\{\int |f_{n-m,\theta} - f_{n,\theta}|\,dx\right\} \le \left(\frac{m}{n-m} + 4\sqrt{\frac{m}{n}}\right)\mathbf{E}\left\{\int |f_{n,\theta} - \mathbf{E}f_{n,\theta}|\,dx\right\}.$$

The result now follows without work. □

Combining Theorems 1 and 2 with Theorem 3.1, we readily obtain the main performance bound of this chapter.

THEOREM 10.3. *Let the set Θ determine a class of regular additive density estimates with $\int f_{n-m,\theta} = 1$ for all $\theta \in \Theta$. Then for all n, $m \le n/2$, Θ, and f:*

$$\mathbf{E}\left\{\int |f_n - f|\right\} \le 3 \inf_{\theta\in\Theta} \mathbf{E}\left\{\int |f_{n,\theta} - f|\right\}\left(1 + \frac{2m}{n-m} + 8\sqrt{\frac{m}{n}}\right) + 8\sqrt{\frac{\log 2\mathbf{S}_{\mathcal{A}_\Theta}(m)}{m}} + \frac{3}{n}.$$

Recalling Exercise 6.2, we note that if the estimates in the class do not satisfy the condition $\int f_{n-m,\theta} = 1$ for all $\theta \in \Theta$, then we still have the same performance bound with slightly worse constants.

COROLLARY 10.1. *Let the set Θ determine a class of regular additive density estimates. Then for all n, $m \le n/2$, Θ, and f:*

$$\mathbf{E}\left\{\int |f_n - f|\right\} \le 5 \inf_{\theta\in\Theta} \mathbf{E}\left\{\int |f_{n,\theta} - f|\right\}\left(1 + \frac{2m}{n-m} + 8\sqrt{\frac{m}{n}}\right) + 8\sqrt{\frac{\log 2\mathbf{S}_{\mathcal{A}_\Theta}(m)}{m}} + \frac{5}{n}.$$

§10.3. Histogram Estimates

The simplest additive estimates are histograms. A histogram density estimate based on a partition P of $\mathbf{R}^d$ is defined by

$$f_{n,P}(x) = \frac{\mu_n(A_P(x))}{\lambda(A_P(x))},$$

where μ_n is the empirical measure based on the sample $X_1, \ldots, X_n$, λ is the Lebesgue measure on $\mathbf{R}^d$, $A_P(x)$ denotes the cell of the partition P into which x falls, and a/∞ is defined to be zero.

As a first application of Theorem 3, we consider the selection of a partition for a histogram. Formally, let $\mathcal{P}$ be a family of partitions of $\mathbf{R}^d$, and to each partition $P \in \mathcal{P}$, assign the corresponding histogram estimate $f_{n,P}(x)$. We use the minimum distance estimate based on data splitting, as described in Section 1, to select a partition from the class $\mathcal{P}$, thus obtaining the density estimate f_n. First observe that any histogram estimate is regular and additive. However, if P contains a cell with infinite Lebesque measure, then $f_{n,P}$ does not necessarily integrate to one, which makes the use of Theorem 3 illegitimate. To fix this problem, we may either use Corollary 1, of first transform $\mathbf{R}^d$ into a compact cube, say $[0,1]^d$, and consider only partitions of the cube.

To apply Theorem 3, we merely need to obtain upper bounds for the shatter coefficients $\mathbb{S}_{\mathcal{A}_\mathcal{P}}(m)$, where $\mathcal{A}_\mathcal{P}$ is the Yatracos class of all sets of the form

$$\left\{x : f_{n-m,P_1}(x) > f_{n-m,P_2}(x)\right\}, \qquad P_1, P_2 \in \mathcal{P}.$$

As a prototype example, consider the class $\mathcal{P}$ of partitions of $\mathbf{R}$ into intervals of length h, where $h = 2^k$ for some integer k. The estimate $f_{n,P}$ based on such a partition is called a regular histogram estimate. To be specific, assume that each partition is anchored at the origin, that is, 0 lies on the boundary of two cells. The parameter k may take infinitely many possible values, and our goal is to select a close-to-optimal value. The following combinatorial argument guarantees, via Theorem 3, that the minimum distance estimate of Section 1 works for the selection of h from a dyadic collection of interval lengths.

LEMMA 10.5.

$$\mathbb{S}_{\mathcal{A}_\mathcal{P}}(m) \leq (m+1)n^2.$$

Also, for all n, $m \leq n/2$, and f, if f_n is the regular histogram estimate picked by the minimum distance method for $\mathcal{P}$, we have

$$\mathbf{E}\left\{\int |f_n - f|\right\} \leq 3 \inf_{\theta\in\mathcal{P}} \mathbf{E}\left\{\int |f_{n,\theta} - f|\right\}\left(1 + \frac{2m}{n-m} + 8\sqrt{\frac{m}{n}}\right) + 8\sqrt{\frac{\log(2(m+1)n^2)}{m}} + \frac{3}{n}.$$

PROOF. Each partition P_k in $\mathcal{P}$ is indexed by a (possibly negative-valued) integer k. Let $A_k(x)$ be the cell of P_k containing x. Observe that the value of the vector

$$z_k = \left(\sum_{i=1}^{n-m} 1_{[y_1 \in A_k(X_i)]}, \ldots, \sum_{i=1}^{n-m} 1_{[y_m \in A_k(X_i)]}\right)$$

can take at most n different values when we vary k but keep the y_i's and X_i's fixed. To see this, observe that there exists a large integer k_0 such that all data points fall in the same cell of P_{k_0}. Then for all $k \geq k_0$ the value of z_k is the same. As k decreases in steps of size one, z_k can only change when there exists a cell $[a,b)$ of P_k such that $[a,(a+b)/2)$ and $[(a+b)/2,b)$ both contain at least one data point among $\{X_1,\ldots,X_{n-m},y_1,\ldots,y_m\}$. But this can happen for at most $n-1$ different values of k. Thus,

$$|\{(z_k,z_\ell) : k,\ell \in \{0,\pm 1,\pm 2,\ldots\}\}| \le n^2.$$

If w and w' are two possible values of the vector z_k, let $U_{w,w'}$ be the set of all pairs (k,ℓ) such that $(z_k,z_\ell) = (w,w')$. Fix $w = (w_1,\ldots,w_m)$ and $w' = (w'_1,\ldots,w'_m)$. If $(k,\ell) \in U_{w,w'}$, then $y_i \in A_{k,\ell} = \{x : f_{n-m,k}(x) > f_{n-m,\ell}(x)\}$ if and only if

$$\frac{w_i}{2^k} > \frac{w'_i}{2^\ell}.$$

Therefore,

$$\begin{aligned}&|\{\{y_1,\ldots,y_m\} \cap A_{k,\ell} : (k,\ell) \in U_{(w,w')}\}| \\ &\quad \le |\{(\mathbf{1}_{[w_1 \ge cw'_1]},\ldots,\mathbf{1}_{[w_m \ge cw'_m]}) : c > 0\}| \le m+1.\end{aligned}$$

Thus, $\mathbb{S}_{\mathcal{A}_\mathcal{P}}(m) \le (m+1)n^2$. □

The exercises contain several other classes of partitions, including various multivariate examples.

§10.4. Bibliographic Remarks

Theorem 2 is a straightforward extension of a slightly less general inequality in Devroye and Lugosi (1996). The best constant in Khintchine's inequality is due to Szarek (1976) (see also Haagerup, 1978). The basis of the proof given here is Lemma 7.1 in Devroye and Györfi (1985). For various results on the regular histogram estimate we refer to Devroye and Györfi (1985). Consistency of the histogram estimate based on data-dependent partitions is investigated in Lugosi and Nobel (1996). Selection of the partition based on penalized maximum likelihood methods is studied by Barron, Birgé, and Massart (1999) and Castellan (2000).

§10.5. Exercises

Exercise 10.1 Consider a fixed class of densities $\mathcal{F} = \mathcal{F}_1 \cup \mathcal{F}_2$, where $\mathcal{F}_1$ is a totally bounded class with covering numbers N_ϵ, and $\mathcal{F}_2$ is a class such that the corresponding Yatracos class $\mathcal{A}$ has Vapnik–Chervonenkis dimension $V < \infty$. Devise a density estimate f_n based on a combination of data splitting with the skeleton estimate and the minimum distance estimate such that for every $f \in \mathcal{F}$, and every $\epsilon > 0$:

$$\mathbf{E}\left\{\int |f_n - f|\right\} \le c\left(\epsilon + \sqrt{\frac{\log N_\epsilon}{n}} + \sqrt{\frac{V}{n}}\right),$$

where c is an absolute constant.

Exercise 10.2 Let f_n be the regular histogram density estimate based on a partition of $\mathbf{R}$ into cells of equal length $h > 0$. Show that if $h \to 0$ and $nh \to \infty$, then for all densities f:

$$\mathbf{E}\{\int |f_n - f|\} \to 0.$$

Conclude that the minimum distance estimate defined in Section 3 is universally consistent. Use the bounded difference inequality to show that, under the same conditions, $\int |f_n - f| \to 0$ a.s. HINT: Proceed as in the proof of the consistency of the kernel estimate.

Exercise 10.3 Let the family $\mathcal{P}$ contain all partitions of $\mathbf{R}$ into $k+1$ intervals. Show that

$$\mathbb{S}_{\mathcal{A}_\mathcal{P}}(m) \leq 2^{4k+1}(n+2k)^{2k}.$$

How large can k be such that there exists an m for the minimum distance method that guarantees that the last two terms in the upper bound of Theorem 3 are $o(n^{-1/3})$?

Exercise 10.4 Let $\mathcal{P}$ contain all partitions (arrangements) of $\mathbf{R}^d$ obtained by k hyperplane cuts. Show that

$$\mathbb{S}_{\mathcal{A}_\mathcal{P}}(m) \leq \left(2\sum_{i=0}^{d}\binom{n-1}{i}\right)^{2k}\binom{2k}{k}2^{\sum_{i=0}^{d}\binom{2k}{i}} \leq n^{kd}2^{2k}2^{(2k)^d}.$$

HINT: First observe that n points by $2k$ hyperplanes can be partitioned at most $\left(2\sum_{i=0}^{d}\binom{n-1}{i}\right)^{2k}$ different ways.

Exercise 10.5 A $k \times \cdots \times k$ flexible grid partition of $\mathbf{R}^d$ is defined by partitioning each of the coordinate axes into k intervals of arbitrary length, and each cell is defined as a Cartesian product of these intervals, leading thus to k^d cells. Let $\mathcal{P}$ contain all $k \times \cdots \times k$ flexible-grid partitions of $\mathbf{R}^d$. Prove that

$$\mathbb{S}_{\mathcal{A}_\mathcal{P}}(m) \leq \left[\binom{n+2k}{2k}\binom{2k}{k}\right]^d 2^{(2k+1)^d}.$$

Exercise 10.6 Let P_0 be a partition of $\mathbf{R}$ into intervals of length $h > 0$, and define $\mathcal{P} = \{P_a : a \in [0,h)\}$ the class of partitions obtained by shifting the anchor point of P_0, where P_a denotes the partition obtained by adding a to each cell-border of P_0. Show that

$$\mathbb{S}_{\mathcal{A}_\mathcal{P}}(m) \leq n^2.$$

Exercise 10.7 Let $q \geq 1$ be an integer, and let $\mathcal{P}_q$ be the class of partitions of $\mathbf{R}$ (anchored at 0) into intervals of length h, where $h = 2^{k+p/q}$ for some integer k and $p \in \{0, 1, \ldots, q-1\}$. Show that

$$\mathrm{S}_{\mathcal{A}_{\mathcal{P}_q}}(m) \leq (m+1)q^2 n^2.$$

Exercise 10.8 Let $q \geq 1$ be an integer, and let $\mathcal{P}'_q$ be the class of partitions of $\mathbf{R}$ (anchored arbitrarily) into intervals of length h, where $h = 2^{k+p/q}$ for some integer k and $p \in \{0, 1, \ldots, q-1\}$. Show that

$$\mathrm{S}_{\mathcal{A}_{\mathcal{P}'_q}}(m) \leq (m+1)q^2 n^4.$$

Exercise 10.9 Let $\sup_y \int |K(x,y)|\, dx < \infty$, where $K(x,y)$ is the kernel in the definition of a regular additive density estimate g_n. Obtain an exponential tail bound for $\mathbf{P}\{|\int |g_n - f| - \mathbf{E}\{\int |g_n - f|\}| > t\}$.

§10.6. References

A. Barron, L. Birgé, and P. Massart, "Risk bounds for model selection via penalization," *Probability Theory and Related Fields*, vol. 113, pp. 301–415, 1999.

G. Castellan, "Sélection d'histogrammes ou de modèles exponentiels de polynomes par morceaux à l'aide d'un critère de type Akaike," Thèse, Mathématiques, Université de Paris-Sud, 2000.

L. Devroye and L. Györfi, *Nonparametric Density Estimation: The L_1 View*, Wiley, New York, 1985.

L. Devroye and G. Lugosi, "A universally acceptable smoothing factor for kernel density estimates," *Annals of Statistics*, vol. 24, pp. 2499–2512, 1996.

U. Haagerup, "Les meilleures constantes de l'inégalité de Khintchine," *Comptes Rendus des Séances de l'Académie des Sciences de Paris. Séries A*, vol. 286, pp. 259–262, 1978.

G. Lugosi and A. Nobel, "Consistency of data-driven histogram methods for density estimation and classification," *Annals of Statistics*, vol. 24, pp. 687–706, 1996.

S. J. Szarek, "On the best constants in the Khintchine inequality," *Studia Mathematica*, vol. 63, pp. 197–208, 1976.

Chapter 11
Bandwidth Selection for Kernel Estimates

§11.1. The Kernel Estimate with Riemann Kernel

This chapter is about the choice of the bandwidth (or smoothing factor) $h \in (0, \infty)$ of the standard kernel estimate

$$f_{n,h}(x) = \frac{1}{nh^d} \sum_{i=1}^{n} K\left(\frac{x - X_i}{h}\right).$$

We assume that $K : \mathbf{R}^d \to \mathbf{R}$ is a fixed kernel with $\int K = 1$. We study the minimum distance estimate based on data-splitting as described in the previous chapter, that is, we fix $m \leq n/2$, define a class of densities $\mathcal{F} = \{f_{n-m,h} : h > 0\}$, and project the empirical measure μ_m (defined on the held-out data $X_{m+1}, \ldots, X_n$) on $\mathcal{F}$ by the minimum distance method of Section 6.8. Clearly, every estimate in $\mathcal{F}$ is additive, so if, in addition, K is a bounded function, then Theorem 10.3 applies. To apply the bound of this theorem, we need to compute useful upper bounds for the shatter coefficient $\mathbf{S}_{\mathcal{A}}(m)$ of the class

$$\mathcal{A} \stackrel{\text{def}}{=} \{\{x : f_{n-m,h}(x) > f_{n-m,h'}(x)\} : \ h, h' > 0\}.$$

Of course, this depends on n, m, and the kernel function K. Obtaining meaningful (i.e., polynomial in m and n) upper bounds may be difficult, or even impossible for general kernels (see Exercise 1). Thus, we begin our study with estimates based on simple kernels.

We consider kernels of the form

$$K(x) = \sum_{i=1}^{k} \alpha_i \mathbf{1}_{A_i}(x),$$

where $\mathbf{1}_A$ denotes the indicator function of a set A, $k < \infty$, $\alpha_1, \ldots, \alpha_k \in \mathbf{R}$, and $A_1, \ldots, A_k$ are Borel sets in $\mathbf{R}^d$ with the following property: the intersection of an infinite ray $\{x : x = tx_0, t \geq 0\}$, anchored at the origin, with any A_i is an interval.

Observe that all convex sets and all star-shaped sets satisfy the above requirement. (A set A is called star-shaped if $x \in A$ implies $\lambda x \in A$ for all

$\lambda \in [0,1]$.) The A_i's need not be disjoint. Kernels of this type are called *Riemann kernels* of parameter k. Important special cases include the uniform densities on ellipsoids, balls, and hypercubes. The key combinatorial result is the following:

LEMMA 11.1. *Let* $K = \sum_{i=1}^{k} \alpha_i \mathbf{1}_{A_i}$ *be a Riemann kernel of parameter* k. *For all* $m \geq 1$ *we have*

$$\mathbb{S}_{\mathcal{A}}(m) \leq (m+1)(1+2km(n-m))^2.$$

PROOF. Let $r = n - m$. Define the vector

$$z_u = \left(\sum_{i=1}^{r} K\left(\frac{y_1 - X_i}{u}\right), \ldots, \sum_{i=1}^{r} K\left(\frac{y_m - X_i}{u}\right) \right) \in \mathbf{R}^m.$$

As we increase u from zero to infinity, each component of z_u changes every time $(y_j - X_i)/u$ enters or leaves a set A_q, $1 \leq q \leq k$ for some X_i, $1 \leq i \leq r$. Note that for fixed $(y_j - X_i)$, the evolution is along an infinite ray anchored at the origin. By our assumption on the possible form of the sets A_q, the number of different values a component can take in its history (as $u \uparrow \infty$) is bounded by $2kr$. As there are m components, the cardinality of the set of different values of z_u is bounded as

$$|\{z_u : u > 0\}| \leq 1 + 2kmr.$$

Thus,

$$|\{(z_u, z_v) : u, v > 0\}| \leq (1 + 2kmr)^2.$$

Let $\mathcal{W} = \{(w, w') : (w, w') = (z_u, z_v) \text{ for some } u, v > 0\}$. For fixed $(w, w') \in \mathcal{W}$, let $U_{(w,w')}$ denote the collection of all (u, v) such that $(z_u, z_v) = (w, w')$. For $(u, v) \in U_{(w,w')}$, we have

$$y_i \in A_{u,v} \quad \text{if and only if} \quad w_i \geq \left(\frac{u}{v}\right)^d w'_i,$$

where w, w' have components w_i, w'_i, respectively, $1 \leq i \leq m$. Thus,

$$\begin{aligned} &|\{\{y_1, \ldots, y_m\} \cap A_{u,v} : (u, v) \in U_{(w,w')}\}| \\ &\quad \leq \left| \left\{ \left(\mathbf{1}_{[w_1 \geq c w'_1]}, \ldots, \mathbf{1}_{[w_m \geq c w'_m]} \right) : c \geq 0 \right\} \right| \leq m + 1. \end{aligned}$$

But then

$$\begin{aligned} &\left|\{\{y_1, \ldots, y_m\} \cap A_{u,v} : (u, v) > 0\}\right| \\ &\quad \leq (m+1)|U_{(w,w')}| \leq (m+1)(1+2kmr)^2. \quad \square \end{aligned}$$

For Riemann kernels, Lemma 1 is the last missing link. Plugging the upper bound of the lemma into Theorem 10.3, we obtain a nonasymptotic, density-free inequality. We defer the discussion of the result to the next section, where we deal with general kernels.

§11.2. General Kernels, Kernel Complexity

Of course, most kernel functions used in practice are not Riemann kernels, and bounding the complexity of the Yatracos class based on such kernels is difficult. Luckily, most kernels can be well approximated by Riemann kernels, and this suggests the following solution for the bandwidth selection problem: first select a positive integer k and a Riemann kernel $K' = \sum_{i=1}^{k} \alpha_i \mathbf{1}_{A_i}$ such that

$$\int |K - K'| \le \frac{1}{n}.$$

Note that this is always possible if K is Riemann integrable. Now define the kernel estimates using the approximating kernel

$$f'_{n-m,h}(x) = \frac{1}{n-m} \sum_{i=1}^{n-m} K'_h(x - X_i)$$

for all $h > 0$. Finally, select the smoothing factor H by the minimum distance estimate over this class, and define the density estimate

$$f_n(x) = \frac{1}{n-m} \sum_{i=1}^{n-m} K_H(x - X_i).$$

This H is called the Riemann approximation bandwidth and f_n is the Riemannian kernel estimate. The size of the smallest k for which this is possible depends on the kernel K. We call this the *kernel complexity* κ_n of K:

$$\kappa_n = \min\left\{ k : \exists\, K' = \sum_{i=1}^{k} \alpha_i \mathbf{1}_{A_i} \text{ such that } \int |K - K'| \le \frac{1}{n} \right\}.$$

The main result of this chapter is the following performance bound for f_n:

THEOREM 11.1. *Let K be a bounded kernel with kernel complexity κ_n, and let $m \le n/2$. Then for all densities f, the Riemannian kernel estimate f_n satisfies*

$$\mathbf{E}\left\{ \int |f_n - f| \right\} \le 3\left(1 + \frac{2m}{n-m} + 8\sqrt{\frac{m}{n}}\right) \inf_h \mathbf{E}\left\{ \int |f_{n,h} - f| \right\} + 8\sqrt{\frac{\log\left(2(m+1)(1 + 2\kappa_n m(n-m))^2\right)}{m}} + \frac{31}{n}.$$

PROOF. Observe that for each h, $\int |f_{n-m,h} - f'_{n-m,h}| \le \int |K - K'|$. Thus,

$$
\begin{aligned}
&\mathbf{E}\left\{\int |f_n - f|\right\} \\
&\le \mathbf{E}\left\{\int |f'_{n-m,H} - f|\right\} + \int |K - K'| \\
&\le 3\left(1 + \frac{2m}{n-m} + 8\sqrt{\frac{m}{n}}\right) \inf_h \mathbf{E}\left\{\int |f'_{n,h} - f|\right\} \\
&\quad + 8\sqrt{\frac{\log\left(2(m+1)(1+2\kappa_n m(n-m))^2\right)}{m}} + \int |K - K'| + \frac{3}{n} \\
&\quad \text{(by the results of the previous section)} \\
&\le 3\left(1 + \frac{2m}{n-m} + 8\sqrt{\frac{m}{n}}\right) \inf_h \mathbf{E}\left\{\int |f_{n,h} - f|\right\} \\
&\quad + 8\sqrt{\frac{\log\left(2(m+1)(1+2\kappa_n m(n-m))^2\right)}{m}} \\
&\quad + \left(1 + 3\left(1 + \frac{2m}{n-m} + 8\sqrt{\frac{m}{n}}\right)\right) \int |K - K'| + \frac{3}{n}. \quad \square
\end{aligned}
$$

To understand the implications of Theorem 1, consider the simplest, though suboptimal choice $m = \lfloor n/2 \rfloor$. Then we obtain

$$
\mathbf{E}\left\{\int |f_n - f|\right\} \le 43 \inf_h \mathbf{E}\left\{\int |f_{nh} - f|\right\} + c\sqrt{\frac{\log(n\kappa_n)}{n}} + \frac{31}{n},
$$

where c is a universal constant, independent of f and K. Taking m as a smaller fraction of n, we may decrease the factor 43 to close to 3, at the expense of increasing the factor c in front of the second term. As we will see later, for most kernels the first term converges to zero much slower than $\sqrt{\log n/n}$ for *all* densities. Therefore, the first term asymptotically dominates the second one if the kernel complexity κ_n is a polynomial function of n. In the next section we show that this is the case for all important kernels.

§11.3. Kernel Complexity: Univariate Examples

In this section we provide a list of examples of kernels on $\mathbf{R}$ whose complexity κ_n is bounded by a polynomial of n. Let us warm up with two toy examples:

UNIFORM KERNELS. If $K(x) = \mathbf{1}_A(x)$ for a star-shaped set A, then obviously $\kappa_n = 1$ for all $n > 1$.

ISOSCELES TRIANGULAR DENSITY. If $K(x) = (1 - |x|)_+$, then elementary calculation shows that for all n, $\kappa_n \le n + 1$.

Most important kernels are nonnegative, unimodal, and symmetric about the origin. For such kernels, the following bound is useful.

SYMMETRIC UNIMODAL KERNELS. Consider a symmetric unimodal density (i.e., $K \geq 0$ and $\int K = 1$) on the real line. Let β be the last positive value for which $\int_\beta^\infty K \leq 1/(4n)$. Partition $[0, \beta]$ and $[-\beta, 0]$ into $N = \lceil 4nK(0)\beta \rceil$ equal intervals. On each interval, let K' be constant with value equal to the average of K over that interval. Let $\gamma = \int_\beta^\infty K/K(\beta)$, and set $K'(x) = K(\beta)$ on $[\beta, \beta + \gamma]$ and $[-\beta - \gamma, -\beta]$. Note that $\int K' = 1$, $\int |K - K'| \leq 1/n$, and that K' is Riemann with parameter $k \leq 2N + 2 \leq 8nK(0)\beta + 10$. Thus, $\kappa_n \leq 8nK(0)\beta + 10$.

> EXAMPLE 1 (BOUNDED COMPACT SUPPORT KERNELS). If $K(x) \leq a 1_{[-b,b]}(x)$ and K is symmetric, nonnegative, and unimodal (such as the Epanechnikov kernel), then $\kappa_n \leq 8nab + 10$.

> EXAMPLE 2 (THE NORMAL KERNEL). When $K(x) = e^{-x^2/2}/\sqrt{2\pi}$, we have $K(0) = 1/\sqrt{2\pi}$. Since for $\beta \geq 1$,
>
> $$\int_\beta^\infty \frac{1}{\sqrt{2\pi}} e^{-x^2/2}\, dx \leq \frac{1}{\sqrt{2\pi}} \frac{1}{\beta} e^{-\beta^2/2} \leq \frac{1}{\sqrt{2\pi}} e^{-\beta^2/2},$$
>
> we may take $\beta = \sqrt{2 \log(4n/\sqrt{2\pi})}$. Thus, for all $n > 1$:
>
> $$\kappa_n \leq \frac{8n\sqrt{\log n}}{\sqrt{\pi}} + 10.$$

> EXAMPLE 3 (THE CAUCHY KERNEL). Take $K(x) = 1/(\pi(1+x^2))$. Note that $K(0) = 1/\pi$, and that $\beta = \pi/(4n)$ will do. Therefore,
>
> $$\kappa_n \leq \frac{32n^2}{\pi^2} + 10.$$

> EXAMPLE 4 (KERNELS WITH POLYNOMIAL TAILS0. Note that if K is a symmetric unimodal density, and $|K(x)| \leq c/(1 + |x|^{\gamma+1})$ for some $c < \infty$, $\gamma > 0$, then $\kappa_n = O(n^{1+1/\gamma})$. In fact, for most cases of interest, $\kappa_n = O(n^\alpha)$ for some finite constant $\alpha > 0$.

Sometimes it may be beneficial to use kernels which may take negative values (see Chapters 16 and 17). The next bound will be useful even for kernels that oscillate infinitely many times.

LIPSCHITZ KERNELS WITH A POSSIBLY HEAVY TAIL. Let K be a univariate kernel that is Lipschitz with Lipschitz constant C, and assume that $|K(x)| \leq D/x^2$ for another constant D. Then

$$\kappa_n \leq 1 + 32CD^2n^3.$$

PROOF. Take $r = 4Dn$ and note that $\int_{|x|>r} |K| \le 2D/r = 1/(2n)$. Partition $[-r, r]$ into q equal intervals of length $2r/q$ each. Define a Riemann kernel K' of order q taking a constant value on each of these intervals, equal to the average of K over the intervals. By the Lipschitz condition, on any such interval A, $\int_A |K - K'| \le C(2r/q)^2/2 = 2Cr^2/q^2$. Let $\gamma^+ = (1/K(r)) \int_r^\infty K$ and $\gamma^- = (1/K(-r)) \int_{-\infty}^{-r} K$, and set $K'(x) = K(r)$ on $[r, r+\gamma^+]$ and $K'(x) = K(-r)$ on $[-r-\gamma^-, r]$. Thus, $\int K' = 1$ and

$$\int |K - K'| \le \frac{1}{2n} + \frac{2Cr^2}{q} \le \frac{1}{n}$$

provided that $q \ge 2Cr^2 n = 32CD^2n^3$. □

Finally, we mention a huge class of kernels, containing nearly every one-dimensional kernel.

KERNELS OF BOUNDED VARIATION. If K is symmetric and a difference of two monotone functions, that is, $K = K_1 - K_2$, $K_1 \downarrow 0$, $K_2 \downarrow 0$ on $[0, \infty)$, then each K_1, K_2 may be approximated as above. Thus, in particular, if K is of *bounded variation*, and $|K(x)| \le c/(1 + |x|^{\gamma+1})$ for some $c < \infty$, $\gamma > 0$, then we may approximate with $\kappa_n = O(n^{1+1/\gamma})$.

§11.4. Kernel Complexity: Multivariate Kernels

PRODUCT KERNELS. If $K = K_1 \times \cdots \times K_d$ is a product of d univariate kernels, and if we approximate K_i with K_i' with parameter $\kappa_{nd}^{(i)}$ for all i (where $\kappa_{nd}^{(i)}$ is the kernel complexity of K_i of precision $1/(nd)$), and form $K' = K_1' \times \cdots \times K_d'$, then K' is a weighted sum of indicators of product sets, and it is Riemann with parameter not exceeding $\prod_{i=1}^d \kappa_{nd}^{(i)}$. Furthermore,

$$\begin{aligned}\int |K - K'| &\le \int |K_1 \times \cdots \times K_{d-1} \times K_d - K_1 \times \cdots \times K_{d-1} \times K_d'| \\ &\quad + \ldots \\ &\quad + \int |K_1 \times K_2' \cdots \times K_d' - K_1' \times K_2' \cdots \times K_d'| \\ &\le d\left(\frac{1}{nd}\right) \\ &= \frac{1}{n}.\end{aligned}$$

Thus, κ_n is bounded by $\prod_{i=1}^d \kappa_{nd}^{(i)}$.

KERNELS THAT ARE FUNCTIONS OF $\|x\|$. Assume that $K(x) = M(\|x\|)$, where M is a bounded nonnegative monotone decreasing function on $[0, \infty)$. Then we may approximate M by a stepwise constant function M', and use the Riemann kernel $K'(x) = M'(\|x\|)$ in the estimate as an approximation of K. Clearly,

$$\int |K(x) - K'(x)|\, dx = \int_0^\infty c_d u^{d-1} |M(u) - M'(u)|\, du,$$

where c_d is d times the volume of the unit ball in $\mathbf{R}^d$. We may define M' as follows. Let β be the largest positive number for which $\int_\beta^\infty c_d u^{d-1} M(u)\, du \le 1/(2n)$. Partition $[0, \beta]$ into $N = \lceil 2nc_d M(0)\beta^d \rceil$ equal intervals. On each interval, let M' equal to the average of M over that interval. Let $\gamma = \int_\beta^\infty c_d u^{d-1} M(u)\, du / M(\beta)$, and set $M'(u) = M(\beta)$ on $u \in [\beta, \beta + \gamma]$, and let $M'(u) = 0$ for $u > \gamma$. Clearly $\int K' = 1$, and that K' is Riemann with parameter $k = N + 1 \le 2nc_d K(0)\beta^d + 2$. Moreover,

$$\begin{aligned}
\int |K(x) - K'(x)|\, dx \\
&= \int_0^\beta c_d u^{d-1} |M(u) - M'(u)|\, du + \int_\beta^\infty c_d u^{d-1} |M(u) - M'(u)|\, du \\
&\le \frac{1}{2n} + c_d \beta^{d-1} \int_0^\beta |M(u) - M'(u)|\, du \\
&\le \frac{1}{2n} + c_d \beta^{d-1} \frac{M(0)\beta}{N} \\
&\le \frac{1}{n}.
\end{aligned}$$

Thus,

$$\kappa_n \le 2nc_d M(0)\beta^d + 2.$$

THE MULTIVARIATE STANDARD NORMAL KERNEL. We may apply the bound of the previous paragraph to the multivariate normal density. First note that it suffices to take $\beta = 2\sqrt{2 \log n}$. From this, we deduce that the kernel complexity is

$$\kappa_n = O(n \log^{d/2} n).$$

§11.5. Asymptotic Optimality

One important corollary of Theorem 1 is that asymptotically the error of the estimate stays within a factor of three of that of the kernel estimate with the best possible smoothing factor.

THEOREM 11.2. *Let K be a bounded nonnegative kernel on the real line with complexity κ_n bounded by some polynomial of n. If $m/n \to 0$ and $m/(n^{4/5}\log n) \to \infty$ as $n \to \infty$, then for all densities f on* $\mathbf{R}$*:*

$$\sup_f \limsup_{n\to\infty} \frac{\mathbf{E}\{\int |f_n - f|\}}{\inf_h \mathbf{E}\{\int |f_{n,h} - f|\}} \le 3.$$

The statement is an easy consequence of the fact that by Theorem 1, with the given choice of m,

$$\mathbf{E}\left\{\int |f_n - f|\right\} \le (3 + o(1)) \inf_h \mathbf{E}\left\{\int |f_{nh} - f|\right\} + o(n^{-2/5}),$$

and the following lower bound due to Devroye and Penrod (1984).

LEMMA 11.2. *Let K be a nonnegative kernel on the real line. Then for any density f:*

$$\liminf_{n\to\infty} n^{2/5} \inf_h \mathbf{E}\left\{\int |f_{nh} - f|\right\} \ge 0.86.$$

The kernel estimate with a kernel which may take negative values may have a faster rate of convergence for some densities. This phenomenon is discussed in detail in Chapters 16 and 17.

§11.6. Bibliographic Remarks

The material of this chapter is based on Devroye and Lugosi (1997). Lemma 2 is due to Devroye and Penrod (1984). The first estimate with the universal property given in Theorem 2 was exhibited by Devroye and Lugosi (1996). Sánchez-Sellero and de Uña (Devroye, 1997) report good experimental results if all data are used and not just the first $n - m$ data points. For a practical implementation and experimental comparison, we refer to Devroye (1997).

§11.7. Exercises

Exercise 11.1 Exhibit a kernel function K such that the class of sets

$$\mathcal{A} = \{\{x : f_{1,h}(x) > f_{1,h'}(x)\},\ h, h' \in (0,\infty)\}$$

has infinite Vapnik–Chervonenkis dimension for some data point X_1. HINT: Reading Section 25.5 in Devroye, Györfi, and Lugosi (1996) may help.

Exercise 11.2 Construct a nonnegative kernel K on the real line with $\int K = 1$ such that the kernel complexity κ_n of K is larger than 2^{2^n} for all sufficiently large n.

Exercise 11.3 Consider an ellipsoidal kernel K on $\mathbf{R}^d$ with $\int K = 1$ of the form $K(x) = M(x^T \Sigma^{-1} x)$, where M is a bounded monotone decreasing function on $[0,1]$, and Σ is a positive definite symmetric $d \times d$ matrix. Show that the kernel complexity $\kappa_n \leq cn$ for some constant c.

Exercise 11.4 Let K be a nonnegative kernel supported in $[0,1]$, satisfying the Lipschitz property $|K(x) - K(y)| \leq C|x - y|$. Consider the finite class of kernel estimates $\{f_{n-m,h} : h \in \Theta\}$, where Θ is the set of all $h_i = h_0(1 + 1/\sqrt{n})^i$, $i = 0, 1, \ldots, 2n^{3/2}$, where $h_0 = e^{-n}$. Choose the bandwidth H using the data splitting minimum distance estimate of Section 10.1. Show that if $m/n \to 0$ and $m/(n^{4/5} \log n) \to \infty$ as $n \to \infty$, then for all densities f on $\mathbf{R}$:

$$\sup_f \limsup_{n \to \infty} \frac{\mathbf{E}\{\int |f_{n-m,H} - f|\}}{\inf_h \mathbf{E}\{\int |f_{n,h} - f|\}} \leq 3.$$

HINT: Use Exercise 9.4.

Exercise 11.5 Consider the data-dependent bandwidth H of the previous exercise. Show that

$$\mathbf{E}\left\{\int |f_{n,H} - f|\right\} \leq \mathbf{E}\left\{\int |f_{n-m,H} - f|\right\} + O\left(\frac{\log n}{n}\right),$$

which means that re-using the held out m samples in the final kernel estimate does not effect asymptotic optimality. HINT: Use the bounded difference inequality.

Exercise 11.6 Theorem 1 applies to the Riemannian kernel estimate. When K is the Gaussian kernel on $\mathbf{R}$, and f_n is obtained by the minimum distance method without using any Riemann approximations, and with $\mathcal{A}$ as in Section 1, is $\mathbf{S}_{\mathcal{A}}(m)$ polynomial in n and m? If so, give a bound to replace that of Theorem 1.

§11.8. References

L. Devroye, "Universal smoothing factor selection in density estimation: Theory and practice (with discussion)," *Test*, vol. 6, pp. 223–320, 1997.

L. Devroye, L. Györfi, and G. Lugosi, *A Probabilistic Theory of Pattern Recognition*, Springer-Verlag, New York, 1996.

L. Devroye and G. Lugosi, "A universally acceptable smoothing factor for kernel density estimation," *Annals of Statistics*, vol. 24, pp. 2499–2512, 1996.

L. Devroye and G. Lugosi, "Non-asymptotic universal smoothing factors, kernel complexity and Yatracos classes," *Annals of Statistics*, vol. 25, pp. 2626–2637, 1997.

L. Devroye and C. S. Penrod, "Distribution-free lower bounds in density estimation," *Annals of Statistics*, vol. 12, pp. 1250–1262, 1984.

Chapter 12
Multiparameter Kernel Estimates

In this chapter we illustrate how the minimum distance estimate described in Chapter 10 may be used to select various parameters simultaneously in an almost optimal manner. The examples are all simple multiparameter versions of the kernel estimate. Once again, the methods applied here are fully combinatorial, as the only thing we need in each case is a suitable upper bound for the shatter coefficient appearing in Theorem 10.3.

§12.1. Multivariate Kernel Estimates—Product Kernels

Let

$$\Theta = \{\theta = (h_1, \ldots, h_d) : \min_i h_i > 0\}$$

be the parameter space of all bandwidths. Consider the kernel estimate on $\mathbf{R}^d$ defined by

$$f_{n,\theta}(x) = \frac{1}{n} \sum_{i=1}^{n} K_\theta(x - X_i),$$

where

$$K_\theta(x) = \prod_{j=1}^{d} \frac{1}{h_j} K_j \left(\frac{x^{(j)}}{h_j} \right),$$

$K_1, \ldots, K_d$ are fixed one-dimensional kernels integrating to one, and $x^{(j)}$ is the jth component of x. In this estimate, the smoothing factor varies in each direction. The issue here is the data-based choice of the smoothing factors. We consider only the simplest possible kernels. The estimates f_n referred to in this chapter are all obtained by the data splitting and minimization method described in Chapter 10, with performance governed by the inequality of Theorem 10.3. The only thing we need to do in this chapter, then, is to bound the shatter coefficient of the appropriate Yatracos class $\mathcal{A}_\Theta$.

LEMMA 12.1. *Assume that* $K_j = [-1/2, 1/2]$, $1 \le j \le d$. *Then for* $m(n-m) \ge 2d$ *we have*

$$\mathbb{S}_{\mathcal{A}_\Theta}(m) \le (m+1)\left(\frac{mne}{2d}\right)^{4d}.$$

Lemma 1 shows that the shatter coefficient is polynomial in n and m. In particular, Theorem 10.3 implies bounds of the form

$$\mathbf{E}\left\{\int |f_n - f|\right\} \le c_0 \inf_{\theta\in\Theta} \mathbf{E}\left\{\int |f_{n,\theta} - f|\right\} + c_1\sqrt{\frac{d\log n}{m}},$$

where c_0 and c_1 are universal constants whose value is easy to determine from Theorem 10.3 and the lemma above, depending on the choice of m. The bounds are as in the one-dimensional case. However, $\sqrt{d}$ factor in the last term of the upper bound, which is typically much smaller than the first term. It is remarkable that adjusting d parameters is not appreciably more difficult than adjusting one parameter.

For general products of Riemann kernels, bounds similar to those of Lemma 1 may be obtained. For products of polynomially Riemann approximable kernels, one needs to optimize a criterion that involves the Riemann approximations, just as in the previous chapters. The details are left as exercises.

PROOF OF LEMMA 1. Denote the jth component of y_t by $y_l^{(j)}$, $1 \le t < m$, $1 \le j \le d$, and the jth component of X_i by $X_i^{(j)}$, $1 \le i \le n-m$, $1 \le j \le d$. For each $\theta \in \Theta$ define the vector

$$\begin{aligned} z_\theta &= \left(z_\theta^{(1)}, \ldots, z_\theta^{(m)}\right) \\ &= \left(\sum_{i=1}^{n-m}\prod_{j=1}^{d} K_j\left(\frac{y_1^{(j)} - X_i^{(j)}}{h_j}\right), \ldots, \sum_{i=1}^{n-m}\prod_{j=1}^{d} K_j\left(\frac{y_m^{(j)} - X_i^{(j)}}{h_j}\right)\right). \end{aligned}$$

Observe that for each $t \le m$ and $i \le n-m$:

$$\prod_{j=1}^{d} K_j\left(\frac{y_t^{(j)} - X_i^{(j)}}{h_j}\right) = 1 \text{ if and only if } \quad y_t - X_i \in R_\theta,$$

where R_θ denotes the rectangle $[-h_1/2, h_1/2] \times \cdots \times [-h_d/2, h_d/2]$. Since there are $m(n-m)$ possible values for $y_t - X_i$, the number of different values the vector z_θ can take as θ varies through Θ is at most $\mathbb{S}_{\mathcal{B}}(m(n-m))$, where $\mathcal{B}$ is the class of all rectangles in $\mathbf{R}^d$. But recall from Lemma 4.1 and Corollary 4.1 that for $m(n-m) \ge 2d$ the shatter coefficient of this class is bounded by $\mathbb{S}_{\mathcal{B}}(m(n-m)) \le (m(n-m)e/2d)^{2d}$. It follows that

$$|\{(z_{\theta_1}, z_{\theta_2}) : \theta_1, \theta_2 \in \Theta\}| \le \left(\frac{m(n-m)e}{2d}\right)^{4d}$$

We finish the proof as in Lemma 11.1. Define

$$\mathcal{W} = \{(w, w') : (w, w') = (z_{\theta_1}, z_{\theta_2}) \text{ for some } \theta_1, \theta_2 \in \Theta\}.$$

For fixed $(w, w') \in \mathcal{W}$, let $U_{(w,w')}$ denote the collection of all (θ_1, θ_2) such that $(z_{\theta_1}, z_{\theta_2}) = (w, w')$. For $(\theta_1, \theta_2) \in U_{(w,w')}$, we have

$$y_t \in A_{\theta_1,\theta_2} \quad \text{if and only if} \quad z_{\theta_1}^{(t)} \prod_{j=1}^{d} \frac{1}{h_{j,1}} \geq z_{\theta_2}^{(t)} \prod_{j=1}^{d} \frac{1}{h_{j,2}},$$

where $\theta_i = (h_{1,i}, \ldots, h_{d,i})$ for $i = 1, 2$. Within the set $U_{(w,w')}$, $z_{\theta_1}^{(t)}$ and $z_{\theta_2}^{(t)}$ are fixed for all t, and therefore

$$\begin{aligned}&|\{\{y_1, \ldots, y_m\} \cap A_{u,v} : (u, v) \in U_{(w,w')}\}| \\ &\qquad \leq \left|\left\{\left(\mathbf{1}_{[w_1 \geq c w_1']}, \ldots, \mathbf{1}_{[w_m \geq c w_m']}\right) : c \geq 0\right\}\right| \leq m + 1,\end{aligned}$$

where $w_1, \ldots, w_m$ and $w_1', \ldots, w_m'$ denote the components of the vectors w and w', respectively. But then

$$\begin{aligned}&|\{\{y_1, \ldots, y_m\} \cap A_{u,v} : (u, v) > 0\}| \\ &\qquad \leq (m+1)|U_{(w,w')}| \leq (m+1)\left(\frac{m(n-m)e}{2d}\right)^{4d}. \quad \square\end{aligned}$$

Let us conclude this section with a few remarks about

$$B_n = \inf_{\theta \in \Theta} \mathbf{E}\left\{\int |f_{n,\theta} - f|\right\}$$

and

$$B_n' = \inf_{\theta \in \Theta'} \mathbf{E}\left\{\int |f_{n,\theta} - f|\right\},$$

where

$$\Theta' = \{\theta = (h_1, \ldots, h_d) : h_1 = \cdots = h_d > 0\}.$$

Theorem 9.2 implies that if $h_1 = \cdots = h_d = h$, $h \to 0$ and $nh^d \to \infty$, then $\mathbf{E}\{\int |f_{n,\theta} - f|\} \to 0$, and thus $B_n' \to 0$. Because $B_n \leq B_n'$, we also have $B_n \to 0$, and the minimum distance estimate f_n is universally consistent provided that $m = o(n)$, $m/\log n \to \infty$. The jump from the same bandwidth in all directions (Θ') to d bandwidths (Θ) is often beneficial as the following example shows. Let $f(x^{(1)}, x^{(2)}) = f_1(x^{(1)}) \times f_2(x^{(2)})$, where f_1 is the exponential density, and f_2 is the normal density. In this example, a bit of work (see the exercises) shows that B_n decreases as $n^{-2/7}$, while B_n' decreases as $n^{-1/4}$. Thus, $B_n'/B_n \to \infty$ at least as fast as $n^{1/28}$.

§12.2. Multivariate Kernel Estimates—Ellipsoidal Kernels

The general multivariate kernel estimate $f_{n,h_1,\ldots,h_d}(x)$ is of the form

$$\frac{1}{n}\sum_{i=1}^{n}\frac{1}{h_1h_2\cdots h_d}K\left(\frac{x^{(1)}-X^{(1)}}{h_1},\frac{x^{(2)}-X^{(2)}}{h_2},\ldots,\frac{x^{(d)}-X^{(d)}}{h_d}\right)$$

for a d-dimensional kernel K. The product kernel of the previous section is merely an important special case. The estimate is universally consistent when $\sum_i h_i \to 0$ and $n\prod_i h_i \to \infty$. Within the boundaries of this condition, the choice of the h_i's is very difficult. Thus, in a multivariate setting, minimization methods become even more important than in the one-dimensional case. In this section, we look at kernels that are affine transformations of radially symmetric kernels. More formally, define the kernel estimate

$$f_{n,\theta}(x) = \frac{1}{n}\sum_{i=1}^{n} K_\theta(x - X_i),$$

where $\theta = \Sigma$, and Σ is a positive definite symmetric $d \times d$ matrix, and

$$K_\theta(x) = v_\theta \mathbf{1}_{[x^T\Sigma^{-1}x\le 1]}.$$

Here v_θ is a normalizing factor such that $\int K_\theta = 1$, and x^T denotes the transpose of the vector x. Observe that here we have significantly more free parameters to set ($d(d+1)/2$ as opposed to d) than in the general multivariate kernel estimate defined above.

LEMMA 12.2. *Let $\mathcal{E}$ be the class of all ellipsoids of $\mathbf{R}^d$, that is, the class of all sets of the form $E_\theta = \{x : x^T\Sigma^{-1}x \le 1\}$. The shatter coefficient of the class is bounded by*

$$\mathbb{S}_{\mathcal{E}}(\ell) \le \left(\frac{\ell e}{d^2/2+d/2+1}\right)^{d^2/2+d/2+1}$$

whenever $\ell \ge d^2+d+2$. The shatter coefficient of the Yatracos class for $f_{n,\theta}(x)$, $\theta \in \Theta$, is bounded by

$$\mathbb{S}_{\mathcal{A}_\Theta}(m) \le (m+1)\left(\frac{m(n-m)e}{d^2/2+d/2+1}\right)^{d^2+d+2}$$

whenever $m(n-m) \ge d^2+d+2$.

PROOF. The former statement follows from Corollary 4.1 since by Corollary 4.2 the Vapnik–Chervonenkis dimension of $\mathcal{E}$ is bounded by $d^2/2 + d/2+1$. For the latter statement, argue as in the proof of Lemma 1, bounding the shatter coefficient by $(m+1)\mathbb{S}^2_{\mathcal{E}}(m(n-m))$. □

Although it is computationally challenging to optimize all $\binom{d}{2}$ entries in a matrix, at least in theory, our minimum distance estimate with an appropriate choice of m guarantees that the expected L_1 error is asymptotically bounded by roughly three times the best possible L_1 error over all such matrices. Again, no conditions are placed on the density or the values of the entries in the matrix. The square root of the number of "free parameters" (which is $d(d+1)/2$ in this case) appears as a multiplicative coefficient in the second term of the upper bound for the L_1 error of the selected estimate. We have already seen this pattern in the previous section. It will reappear in the next section.

As in the univariate case, the argument may be extended via Riemann approximations to the class of estimates with $K_\theta(x) = v_\theta L(x^T\sigma^{-1}x)$, where $L : \mathbf{R}^+ \to \mathbf{R}$ is a fixed function. The details are left as an exercise.

§12.3. Variable Kernel Estimates

Variable kernel estimates use different smoothing factors in different regions of $\mathbf{R}^d$. To obtain good estimates, the smoothing factors and the regions should be selected based on the data. More precisely, if $P = \{B_1, B_2, \ldots, B_r\}$ is a partition of $\mathbf{R}^d$ into r sets, then the corresponding variable kernel estimate is defined as

$$f_{n,\theta}(x) = \frac{1}{n}\sum_{i=1}^{n}\sum_{j=1}^{r} K_{h_j}(x - X_i)\mathbf{1}_{B_j}(X_i),$$

where $\mathbf{h} = (h_1, \ldots, h_r) \in \left(\mathbf{R}^+\right)^r$ is a vector of positive smoothing factors, and $\theta = (P, \mathbf{h})$. Clearly, $\int K = 1$ implies that $\int f_{n,\theta} = 1$.

Such estimates promise substantially improved performance over ordinary kernel estimates for densities with varying behavior in different regions of the space. To exploit the advantages offered by variable kernel estimates, one must design a good data-dependent way of determining the partition and the different smoothing factors in each cell. The minimum distance estimate is an ideal tool for this kind of problem. Of course, the possible partitions among which we can select have to be restricted somehow. Larger families of partitions offer better flexibility, but it is more difficult to select the best among them. The complexity of the class of possible partitions will appear in the bounds.

Formally, let $\mathcal{P}$ be a family of partitions of $\mathbf{R}^d$ such that each $P \in \mathcal{P}$ has at most r cells. Let $\Theta = \{(P, \mathbf{h}) : P \in \mathcal{P}, \mathbf{h} \in \left(\mathbf{R}^+\right)^r\}$ denote the set of all possible pairs of partitions and vectors of smoothing factors. With a slight abuse of notation, denote by $\mathbf{S}_{\mathcal{P}}(n)$ the nth shatter coefficient of the class of sets which are cells of any partition in $\mathcal{P}$. For simplicity, we assume that K is the window kernel $K(x) = c\mathbf{1}_{[\|x\|\le 1]}$, where c is an appropriate normalizing factor.

LEMMA 12.3. *If $\mathcal{A}_\Theta$ is the Yatracos class defined by* $\Theta = \{(P, \mathbf{h}) : P \in \mathcal{P}, \mathbf{h} \in (\mathbf{R}^+)^r\}$, *then*

$$\mathbb{S}_{\mathcal{A}_\Theta}(m) \le 2m^{2r}\left((m(n-m)+1)\mathbb{S}_{\mathcal{P}}(m(n-m))\right)^{2r}.$$

PROOF. For each $\theta \in \Theta$, consider the $m \times r$ matrix z_θ such that the element in its tth row and jth column is

$$z_\theta^{(t,j)} = \sum_{i=1}^{n-m} \mathbf{1}_S\left(\frac{y_t - X_i}{h_j}\right)\mathbf{1}_{B_j}(X_i), \; t \le m, \; j \le r,$$

where $S = \{x : \|x\| \le 1\}$ is the closed unit ball centered at the origin. Clearly,

$$\mathbf{1}_S\left(\frac{y_t - X_i}{h_j}\right)\mathbf{1}_{B_j}(X_i) = 1 \text{ if and only if } (X_i, y_t - X_i) \in B_j \times S(0, h_j),$$

where $S(0,h) = \{x : \|x\| \le h\}$. Since there are $m(n-m)$ different pairs $(X_i, y_t - X_i)$, the number of different values the jth column $(z_\theta^{(1,j)}, \ldots, z_\theta^{(m,j)})$ of the matrix z_θ can take as $\theta \in \Theta$ is at most the shatter coefficient $\mathbb{S}_{\mathcal{C}}(m(n-m))$ of the class of sets $\mathcal{C}$ of the form $B \times S(0,h)$, where h ranges through the set of positive numbers, and B is any set in any possible partition in $\mathcal{P}$. This shatter coefficient is clearly bounded by the product of the $m(n-m)$th shatter coefficient of the class of sets $\{S(0,h) : h > 0\}$ and the shatter coefficient $\mathbb{S}_{\mathcal{P}}(m(n-m))$ of the class of sets which are cells of any Partition in $\mathcal{P}$. Since the $m(n-m)$th shatter coefficient of $\{S(0,h) : h > 0\}$ is $m(n-m)+1$, the jth column of z_θ can take at most $(m(n-m)+1)\mathbb{S}_{\mathcal{P}}(m(n-m))$ values. But since the matrix z_θ has r columns, it can take at most

$$\left((m(n-m)+1)\mathbb{S}_{\mathcal{P}}(m(n-m))\right)^r$$

values. Thus,

$$|\{(z_{\theta_1}, z_{\theta_2}) : \theta_1, \theta_2 \in \Theta\}| \le \left((m(n-m)+1)\mathbb{S}_{\mathcal{P}}(m(n-m))\right)^{2r}.$$

Let $\mathcal{W} = \{(w, w') : (w, w') = (z_{\theta_1}, z_{\theta_2}) \text{ for some } \theta_1, \theta_2 \in \Theta\}$. For fixed $(w, w') \in \mathcal{W}$, let $U_{(w,w')}$ denote the collection of all (θ_1, θ_2) such that $(z_{\theta_1}, z_{\theta_2}) = (w, w')$. For $(\theta_1, \theta_2) \in U_{(w,w')}$, and $t \le m$ we have

$$y_t \in A_{\theta_1,\theta_2} \text{ if and only if } \sum_{j=1}^r \frac{1}{h_{j,1}} z_{\theta_1}^{(t,j)} \ge \sum_{j=1}^r \frac{1}{h_{j,2}} z_{\theta_2}^{(t,j)},$$

where $\theta_1 = (h_{1,1}, \ldots, h_{r,1})$ and $\theta_2 = (h_{1,2}, \ldots, h_{r,2})$. Within the set $U_{(w,w')}$, $z_{\theta_1}^{(t,j)}$ and $z_{\theta_2}^{(t,j)}$ are fixed for all t and j, and therefore the number of different values of the binary vector

$$\left(\mathbf{1}_{A_{\theta_1,\theta_2}}(y_1), \ldots, \mathbf{1}_{A_{\theta_1,\theta_2}}(y_m)\right)$$

is at most the number of different values the vector

$$\left(1_{\left[\sum_{j=1}^{r}\left((1/h_{j,1})z_{\theta_1}^{(1,j)}-(1/h_{j,2})z_{\theta_2}^{(1,j)}\right)\geq 0\right]},\ldots,\right.$$
$$\left.1_{\left[\sum_{j=1}^{r}\left((1/h_{j,1})z_{\theta_1}^{(m,j)}-(1/h_{j,2})z_{\theta_2}^{(m,j)}\right)\geq 0\right]}\right)$$

takes as $h_{1,1},\ldots,h_{r,1},h_{1,2},\ldots,h_{r,2}$ all vary through $\mathbf{R}^+$. But this is not more than the maximal number of different ways of dichotomizing m points in $\mathbf{R}^{2r}$ by hyperplanes, which is at most $2m^{2r}$ (see Chapter 4). The proof of Lemma 3 is finished. □

As a simple but important example, consider $d = 1$, and let $\mathcal{P}$ be the class containing all partitions of the real line into at most r intervals. Then $\mathbf{S}_{\mathcal{P}}(n)$ is just the nth shatter coefficient of the class of all intervals, which equals $n(n+1)/2+1$, and therefore

$$\mathbf{S}_{\mathcal{A}_\Theta}(m) \leq 2m^{2r}(m^2(n-m)^2)^{2r}(2m(n-m))^{2r} \leq 2^{1+2r}m^{8r}n^{6r}.$$

For fixed r, this bound is polynomial in n and m, and the remarks of the previous sections apply. Note that r is allowed to tend to infinity, but should not increase so fast that the shatter coefficient term in the upper bound of Theorem 10.3 starts dominating. It should be of interest to investigate the possibility or impossibility of finding data-based choices for the variable kernel estimate in which each X_i has its own tunable H_i, as in the original estimate of Breiman, Meisel, and Purcell (1977). Unfortunately, as we will see in Section 15.7, this class of estimates is too large, as it is impossible to find any data-based method for selecting an estimate whose error is within a constant factor of that of the best in the class. Hence the need to restrict the complexity of the class of tunable parameters as we did in this section. This theme will also motivate Section 5.

§12.4. Tree-Structured Partitions

A popular way of partitioning $\mathbf{R}^d$ recursively is by tree-structured partitions, starting from a single set, $\mathbf{R}^d$ itself. Refinements are made iteratively by splitting a set of the current partition by a hyperplane orthogonal to a coordinate axis. Such trees are called k-d trees. Let $\mathcal{P}$ denote the class of all tree-structured partitions obtained by $r-1$ splits. Then clearly, each element of $\mathcal{P}$ partitions $\mathbf{R}^d$ into r rectangles. Since the Vapnik–Chervonenkis of the class of rectangles is $2d$ (Lemma 4.1), we have for $\ell \geq 2d$:

$$\mathbf{S}_{\mathcal{P}}(\ell) \leq \left(\frac{\ell e}{2d}\right)^{2d},$$

so substituting into Lemma 3, we obtain, for $m(n-m) \geq 2d$:

$$\mathbf{S}_{\mathcal{A}_\Theta}(m) \leq 2m^{2r}(m(n-m)+1)^{2r}(m(n-m))^{4dr}.$$

Here it is understood once again that the variable kernel estimate of Section 3 is optimized with respect to $\theta = (P, \mathbf{h})$, where P is a tree partition with $r-1$ splits. The exercises offer other examples of classes of partitions of $\mathbf{R}^d$ to which Lemma 3 can be applied.

§12.5. Changepoints and Bump Hunting

We return to the one-dimensional density estimation problem once again, but offer a multiparameter solution to an age-old problem related to the presence of a finite number of discontinuities in f. Consider a finite number (k) of univariate kernel estimates $\xi_{n-m,1}, \ldots, \xi_{n-m,k}$, where the data splitting notation and methodology of Chapter 10 is followed. The ith estimate $\xi_{n-m,i}$ has its own bandwidth h_i, depending upon i and n only. For now, the kernel choices are unimportant. The parameters we introduce describe a partition of the real line into r intervals by $r-1$ thresholds $t_1 < \cdots < t_{r-1}$. These thresholds, appended by the r indices of the choices of kernel estimates for the (ordered) intervals are collected in a $(2r-1)$-dimensional parameter $\theta = (t_1, \ldots, t_{r-1}, \eta_1, \ldots, \eta_r)$. This means that on the ith interval, we use $f_{n,\theta} = \xi_{n-m,\eta_i}$. The space of these ordered thresholds is called Θ. The minimum distance estimate f_n is obtained by optimization over Θ using the Yatracos set $\mathcal{A}_\Theta$. The only theoretical difficulty is caused by the fact that $\int f_{n,\theta} \neq 1$.

LEMMA 12.4. *The shatter coefficient of $\mathcal{A}_\Theta$ satisfies*

$$\mathrm{S}_{\mathcal{A}_\Theta}(m) \leq (m+1)^{2r-2} k^{2r}.$$

Thus, if the kernels used in the estimates $\xi_{n-m,1}, \ldots, \xi_{n-m,k}$ are bounded, then for $1 \leq m \leq n/2$, the minimum distance estimate f_n satisfies

$$\mathbf{E}\left\{\int |f_n - f|\right\} \leq 5 \inf_{\theta\in\Theta} \mathbf{E}\left\{\int |f_{n,\theta} - f|\right\}\left(1 + \frac{2m}{n-m} + 8\sqrt{\frac{m}{n}}\right) + 8\sqrt{\frac{2r\log(k(m+1))}{m}} + \frac{5}{n}.$$

PROOF. Let $f_{n-m,\theta}$ denote the global estimate with $t_1, \ldots, t_{r-1}, \eta_1, \ldots, \eta_r$ "variable." Take m real numbers $y_1, \ldots, y_m$, and look at their memberships in the sets $\{x : f_{n-m,\theta} > f_{n-m,\theta'}\}$, $\theta \neq \theta'$. Call these memberships $z(\theta, \theta') = (z_1(\theta, \theta'), \ldots, z_m(\theta, \theta'))$, where the z_i's are 1 if y_i is in the set and 0 otherwise. Without loss of generality, we may order $y_1, \ldots, y_m$. Replace each t_i by an integer value v_i representing the number of y_j's smaller than t_i. Thus, $0 \leq v_i \leq m$. The membership bit vector $z(\theta, \theta')$ only depends upon the vector $v = (v_1, \ldots, v_{r-1}, \eta_1, \ldots, \eta_r)$ and a similar vector v' for θ'. The number of possible values for $z(\theta, \theta')$ cannot be larger than the number of possible values for (v, v'). Thus,

$$\mathrm{S}_{\mathcal{A}_\Theta}(m) \leq (m+1)^{2r-2} k^{2r}.$$

The last inequality follows by noting that each estimate $f_{n-m,\theta}$ is a regular additive estimate and using Corollary 10.1, dealing with the case that $\int f_{n,\theta} \neq 1$. □

A multivariate generalization is suggested in Exercise 14. Lemma 4 shows how important it is to keep r small. However, the choice of k is less crucial. It is also interesting to note that the choice of the kernel is immaterial. In fact, the k estimates can be quite arbitrary. As the class of estimates over which we optimize contains $\xi_{n-m,1}$, we have universal consistency if $\xi_{n-m,1}$ is universally consistent (hence the importance of at least including one kernel estimate in the collection) and if $m = o(n)$ and $r \log(mk)/m \to 0$. This manner of combining estimates is useful for aesthetic corrections of an estimate. Large tails may require bigger bandwidths, discontinuities in derivatives may require smaller bandwidths, and discontuities in the density itself even smaller ones. In other words, it provides a tool for picking local bandwidths.

Consider the simple example of the estimation of an exponential density. With the ordinary kernel estimate and uniform kernel on $[-1,1]$, the best possible bandwidth is $h = n^{-1/3}$ modulo a constant. The best possible expected L_1 error decreases as $n^{-1/3}$ as well (see the exercises). In fact, this $n^{-1/3}$ rate is a lower bound for any density with such a simple discontinuity (Devroye, 1987). Recall from Chapter 9 that with positive kernels, an expected error of about $n^{-2/5}$ is achievable, but not improvable, under smoothness and tail conditions. Consider two kernel estimates, both with that uniform kernel. Their bandwidths are $h_1 = n^{-5/17}$ (which is very small) and $h_2 = n^{-1/5}$ (which is very large) respectively. Combine them using a one threshold setting ($r = 2$ in Lemma 4), with threshold at $t_1 = n^{-1/5}$. Use the kernel estimate with bandwidth h_1 (and the full sample!) on $(-\infty, t_1)$, and use the other kernel estimate on $[t_1, \infty)$. The expected L_1 error on that right interval is $O(n^{-2/5})$ simply because the kernel estimate on that interval is not aware of the discontinuity—it is out of its reach—and the density is smooth on $(0, \infty)$. The contribution to the expected L_1 error from $(-\infty, t_1]$ is $O(n^{-7/15})$ (see the exercises) and thus asymptotically negligible. If we place t_1 to the left of $n^{-1/5}/2$, then the L_1 error due to $[t_1, \infty)$ is of the order of a whopping $n^{-1/5}$ and starts dominating. It is clear, then, that the optimal threshold will be found near h_2. In other words, the two-region two-kernel estimate method may be used to automatically locate one discontinuity. To isolate a discontinuity from both sides, a three-region two-kernel method suffices. To isolate k discontinuities or infinite peaks, a $(2k+1)$-region two-kernel estimate method suffices. The thresholds will be found at distances roughly $\pm h_2$ from all discontinuities. More importantly, the overall expected error is explicitly governed by Lemma 4. In our example, we have for $r = 1$, with a kernel estimate with the optimal $h = n^{-1/3}$:

$$\inf_\theta \mathbf{E}\left\{\int |f_{n,\theta} - f|\right\} \sim cn^{-1/3}$$

for some constant c, while for $r = 2$ with bandwidths $h_1 = n^{-7/15}$, $h_2 = n^{-1/5}$:

$$\inf_\theta \mathbf{E}\left\{\int |f_{n,\theta} - f|\right\} \sim cn^{-2/5}.$$

If $m = o(n)$, $m/n^{4/5} \to \infty$, the expected L_1 error of the minimum distance estimate is asymptotically not more than three times that optimum. Interestingly, the same statement, with the same rates of convergence, remains valid with the same two kernel estimates if we have a density that has k simple discontinuities and at least $2k+1$ regions, whenever the density is sufficiently smooth on each open interval that has no discontinuity.

We were a bit careless by dropping constants in the definitions of the bandwidths. To do a good job, the bandwidths, estimate assignments, and thresholds should all be free parameters. This, of course, is very feasible, as you will find out in the exercises.

§12.6. Bibliographic Remarks

The variable kernel estimate was first investigated by Breiman, Meisel, and Purcell (1977). Tree partitioning for density estimation is only in its infancy. For example, k-d trees (Bentley, 1975) are widely used in computer science but have found little acceptance in statistics. A notable exception is the early tree-based histogram estimate of Gessaman (1970). Breiman, Friedman, Olshen, and Stone (1984) and Lugosi and Nobel (1996) prove the consistency of tree-based histogram estimates. For other consistency results about data-based partitioning estimates we refer to Chen and Zhao (1987), Gordon and Olshen (1984), Zhao, Krishnaiah, and Chen (1990). The most recent results on the consistency of multivariate multiparameter kernel estimates are in Devroye and Krzyżak (2000). Some of the examples of this chapter also appear in Devroye, Lugosi, and Udina (2000)

§12.7. Exercises

Exercise 12.1 In the notation of Section 1, consider a bivariate density $f(x^{(1)}, x^{(2)}) = f_1(x^{(1)}) \times f_2(x^{(2)})$, where f_1 is the exponential density, and f_2 is the normal density. The objective is to compute the optimal order of convergence to zero of $\mathbf{E}\{\int |f_{n,\theta} - f|\}$ when the two bandwidths in θ are identical and when they are allowed to be different. The kernel is a product kernel as in Section 1.

A. Consider the bias/variation split as in the proof of Theorem 9.5. Show first that the rate of decrease of $\mathbf{E}\{\int |f_{n,\theta} - \mathbf{E}\{f_{n,\theta}\}|\}$ is of the order of
$$\frac{1}{\sqrt{nh_1h_2}}$$
if $nh_1h_2 \to \infty$ and $\max_i h_i \to 0$. HINT: For the upper bound, argue as in the proof of Theorem 9.5. For the lower bound, just look at a rectangle of $\mathbf{R}^2$ on which f is bounded away from zero.

B. Obtain the rate of decrease of $\int |\mathbf{E}\{f_{n,\theta}\} - f|$ as $\max_i h_i \to 0$. To do so, note that $\mathbf{E}\{f_{n,\theta}\}$ is an integral average over a rectangle of dimension $h_1 \times h_2$ centered at x. In the region $[h_1/2, \infty) \times \mathbf{R}$, show that the bias is of the order of $h_1^2 + h_2^2$. In the region $[0, h_1) \times \mathbf{R}$, show that the integrand of the bias is typically of the order of $h_1 + h_2^2$, so that integrated over this region, it is of the order of $h_1^2 + h_1h_2^2$. Conclude that the bias decreases as $h_1 + h_2^2$.

C. Optimizing the sum of bias and variance contributions, show that $B_n = O(n^{-2/7})$, $B_n' = \Omega(n^{-1/4})$, and conclude that $B_n/B_n' = O(n^{-1/28})$.

Exercise 12.2 Consider the multivariate product kernel estimate on $\mathbf{R}^d$ in which the kernel K is a d-fold product of arbitrary one-dimensional kernels. Let the bandwidths $h_1, \ldots, h_d$ satisfy $\lim_{n\to\infty} \max_i h_i = 0$, and $\lim_{n\to\infty} n \prod_{i=1}^d h_i = \infty$. Show that this estimate is L_1 consistent for all densities f. NOTE: this is not a simple exercise, as the h_i's may tend to zero at very different rates.

Exercise 12.3 Consider the class of multiparameter kernel estimates
$$f_{n,\theta}(x) = \frac{1}{n} \sum_{i=1}^{n} K_\theta(x - X_i)$$
on $\mathbf{R}^d$, where $\theta = (h_1, \ldots, h_d)$ is a vector of positive smoothing factors, and
$$K_\theta(x) = \prod_{j=1}^{d} \frac{1}{h_j} K_j \left(\frac{x^{(j)}}{h_j} \right),$$
is a product kernel with one-dimensional component kernels $K_1, \ldots, K_d$ integrating to one. Assume that each K_j ($j \le d$) is of the form
$$K_j = \sum_{i=1}^{k_j} \alpha_{i,j} 1_{A_{i,j}},$$
where each set $A_{i,j}$ is a closed and bounded interval in $\mathbf{R}$. Extend Lemma 1 to the Yatracos class defined by these estimates.

Exercise 12.4 Consider the class of ellipsoidal kernel estimates

$$f_{n,\theta}(x) = \frac{1}{n}\sum_{i=1}^{n} K_\theta(x - X_i),$$

where $\theta = \Sigma$, and Σ is a positive definite symmetric $d \times d$ matrix, and

$$K_\theta(x) = v_\theta L(x^T \Sigma^{-1} x).$$

Here $L = \sum_{i=1}^{k} \alpha_i 1_{A_i}$ is a fixed function defined on $[0, \infty)$, where each A_i is a finite interval, and $v_\theta = 1/\int L(x^T\Sigma^{-1}x)\,dx$. Derive an upper bound for the shatter coefficient of the corresponding Yatracos class.

Exercise 12.5 Let $K_1, \ldots, K_d$ be nonnegative univariate kernels, and consider the multivariate kernel estimate

$$f_{n,h_1,\ldots,h_d}(x) = \frac{1}{n}\sum_{i=1}^{n}\prod_{j=1}^{d}\frac{1}{h_j}K_j\left(\frac{x^{(j)} - X^{(j)}}{h_j}\right),$$

where $h_1, \ldots, h_d$ are smoothing factors. Obtain the best possible lower bound as a function of n for $\inf_f \inf_{h_1,\ldots,h_d,K_1,\ldots,K_d} \mathbf{E}\{\int |f_n - f|\}$. HINT: Is this very different from the one-dimensional case?

Exercise 12.6 Consider the general multivariate kernel estimate with kernel K:

$$f_{n,h_1,\ldots,h_d}(x) = \frac{1}{n}\sum_{i=1}^{n}\frac{1}{h_1 h_2 \cdots h_d}K\left(\frac{x^{(1)} - X^{(1)}}{h_1}, \frac{x^{(2)} - X^{(2)}}{h_2}, \ldots, \frac{x^{(d)} - X^{(d)}}{h_d}\right).$$

A. Show that all lower-dimensional marginals of $f_{n,h_1,\ldots,h_d}$ are also kernel estimates.
B. Prove an inequality between the total variation errors of kernel estimates and their marginals.
C. Conclude that the universal lower bound for

$$\inf_f \inf_{h_1,\ldots,h_d,K} \mathbf{E}\left\{\int |f_n - f|\right\}$$

for nonnegative univariate kernel estimates remains valid for all nonnegative multivariate kernel estimates.
D. Can this bound be attained (modulo a constant) for certain multivariate densities and certain selections of the bandwidths?

Exercise 12.7 Consider the multivariate kernel estimate with kernel K:

$$f_{n,h}(x) = \frac{1}{n}\sum_{i=1}^{n}\frac{1}{h^d}K\left(\frac{x^{(1)} - X^{(1)}}{h}, \frac{x^{(2)} - X^{(2)}}{h}, \ldots, \frac{x^{(d)} - X^{(d)}}{h}\right).$$

Obtain the best possible lower bound as a function of n for

$$\inf_f \inf_{h,K} \mathbf{E}\left\{\int |f_n - f|\right\}$$

when K is nonnegative. Compare your answer with that of the previous two exercises. NOTE: This exercise is really a small research problem.

Exercise 12.8 (BINARY SPACE PARTITION TREES). A binary space partition tree produces a partition of $\mathbf{R}^d$ recursively, starting with the $\mathbf{R}^d$ as a single cell. Cells are split iteratively by cutting each cell of the current partition by any hyperplane. Let $\mathcal{P}$ denote the class of all such partitions P obtained by $r-1$ splits.

A. Bound the shatter coefficient $\mathbb{S}_{\mathcal{P}}(\ell)$.
B. Assume that the space is $[0,1]^d$ and that $X_1, \ldots, X_d$ are drawn from an unknown density f on $[0,1]^d$. Define the density estimate $f_{n,P}$ as for the histogram (divide the average number of points in a cell by its volume). Let f_n be the minimum distance estimate obtained by optimizing over $\mathcal{P}$. Derive an upper bound for $\mathbf{E}\{\int |f_n - f|\}$ based on that of Theorem 10.3. NOTE: The only thing you have to do is to work out the shatter coefficient for the Yatracos class.
C. Repeat Part B for the kernel estimate with fixed uniform kernel on the unit square, but a different smoothing factor h on each set in the partition. Take h identical in all directions.

Exercise 12.9 (ARRANGEMENTS). A collection of r hyperplanes in $\mathbf{R}^d$ defines a partition P, called an arrangement. Let $\mathcal{P}$ denote the class of all such partitions P.

A. Find an upper bound on the number of possible cells in any such partition P.
B. Bound the shatter coefficient $\mathbb{S}_{\mathcal{P}}(\ell)$.
C. Assume that the space is $[0,1]^d$ and that $X_1, \ldots, X_d$ are drawn from an unknown density f on $[0,1]^d$. Define the density estimate $f_{n,P}$ as for the histogram (divide the average number of points in a cell by its volume). Let f_n be the minimum distance estimate obtained by optimizing over $\mathcal{P}$. Derive an upper bound for $\mathbf{E}\{\int |f_n - f|\}$ based on that of Theorem 10.3.

Exercise 12.10 (VORONOI PARTITIONS). A Voronoi partition of $\mathbf{R}^d$ based on r points $x_1, \ldots, x_r \in \mathbf{R}^d$ has r cells, where the ith cell contains all points in $\mathbf{R}^d$ which are closer to x_i than to any other x_j, $j \neq i$. Points on the boundary of cells belong to the cell with smallest index.

A. Derive a bound for $\mathbb{S}_{\mathcal{P}}(\ell)$ if $\mathcal{P}$ is the class of all Voronoi partitions with at most r cells.

B. Assume that the space is $[0,1]^d$ and that $X_1, \ldots, X_d$ are drawn from an unknown density f on $[0,1]^d$. Define the density estimate $f_{n,P}$ as for the histogram (divide the average number of points in a cell by its volume). Let f_n be the minimum distance estimate obtained by optimizing over $\mathcal{P}$. Derive an upper bound for $\mathbf{E}\{\int |f_n - f|\}$ based on that of Theorem 10.3.
C. Repeat Part B for the kernel estimate with fixed uniform kernel on the unit square, but a different smoothing factor h on each set in the partition. Take h identical in all directions.

Exercise 12.11 (THE EXAMPLE OF SECTION 5). In this exercise, you are to work out the details for the example described in Section 5. Let $K = \frac{1}{2}\mathbf{1}_{[-1,1]}$, and let $f(x) = e^{-x}, x > 0$. Let $f_{n,h}$ denote the kernel estimate with bandwidth h and kernel K.

A. Show that $\inf_h \mathbf{E}\{\int |f_{n,h} - f|\} \geq cn^{-1/3}$ for some constant $c > 0$. HINT: Use the bias and variation inequalities of the proof of Theorem 9.5. Bound the bias from below by its contribution on $[0, h/2]$, and bound the variation from below pointwise for every $x > h$ using properties of the binomial distribution.
B. Show that $\inf_h \mathbf{E}\{\int |f_{n,h} - f|\} \leq cn^{-1/3}$ for some constant $c > 0$ by showing that the upper bound is valid for $h = n^{-1/3}$.
C. Show that $\int_h^\infty |f_{n,h} - f| \leq cn^{-2/5}$ for some constant $c > 0$ when $h = n^{-2/5}$.
D. Show that $\int_{-\infty}^h |f_{n,h} - f| \leq cn^{-7/15}$ for some constant $c > 0$ when $h = n^{-7/15}$.
E. Show that for the minimum distance estimate with two regions and two kernel estimates, one with bandwidth $n^{-2/5}$ and one with bandwidth $n^{-7/15}$:

$$\inf_\theta \mathbf{E}\left\{\int |f_{n,\theta} - f|\right\} = O(n^{-2/5}).$$

F. Finally, if f_n is the minimum distance estimate for Part E, show how you should choose m such that

$$\mathbf{E}\left\{\int |f_n - f|\right\} = O(n^{-2/5}).$$

Exercise 12.12 Extend Lemma 4 to the situation in which the k candidate kernel estimates $\xi_{n-m,1}, \ldots, \xi_{n-m,k}$ each have kernel $K = \frac{1}{2}\mathbf{1}_{[-1,1]}$, but the k bandwidths are now free parameters, in addition to the kernel assignments and the choices of the thresholds.

Exercise 12.13 (CONTINUATION OF PREVIOUS TWO EXERCISES). With the method of the previous exercise, show that for the example of the exercise preceding that, with two regions and two kernel estimates,

$$\inf_\theta \mathbf{E}\left\{\int |f_{n,\theta} - f|\right\} = O(n^{-2/5}).$$

If f_n is the minimum distance estimate, show how you should choose m such that

$$\mathbf{E}\left\{\int |f_n - f|\right\} = O(n^{-2/5}).$$

Exercise 12.14 Consider the estimate of Section 5, but now in a more general setting. Let $\xi_{n-m,1}, \ldots, \xi_{n-m,k}$ be density estimates in $\mathbf{R}^d$ and let $\mathcal{P}$ be a class of partitions such that each element of $\mathcal{P}$ partitions $\mathbf{R}^d$ into r cells. For any cell of a fixed partition $P \in \mathcal{P}$, we can pick one of the extimates $\xi_{n-m,i}$ to use in that cell. Consider the class of all such density estimates, indexed by $P \in \mathcal{P}$ and a vector of r integers with values between 1 and k (corresponding to the index of the estimate used in each cell of the partition). Show that the shatter coefficient of the Yatracos class $\mathcal{A}$ corresponding to this class of estimates is bounded by

$$\mathrm{S}_{\mathcal{A}_\Theta}(m) \le \mathrm{S}_{\mathcal{P}}(m)^{2r} k^{2r}.$$

§12.8. References

J. L. Bentley, "Multidimensional binary search trees used for associative searching," *Communications of the Association of Computing Machinery*, vol. 18, pp. 509–517, 1975.

L. Breiman, J. H. Friedman, R. A. Olshen, and C. J. Stone, *Classification and Regression Trees*, Wadsworth International, Belmont, CA, 1984.

L. Breiman, W. Meisel, and E. Purcell, "Variable kernel estimates of multivariate densities," *Technometrics*, vol. 19, pp. 135–144, 1977.

X. R. Chen and L. C. Zhao, "Almost sure L_1-norm convergence for data-based histogram density estimates," *Journal of Multivariate Analysis*, vol. 21, pp. 179–188, 1987.

L. Devroye, *A Course in Density Estimation*, Birkhäuser-Verlag, Boston, 1987.

L. Devroye, G. Lugosi, and F. Udina, "Inequalities for a new data-based method for selecting nonparametric density estimates," in: *Festschrift in Honour of George Roussas* (edited by M. L. Puri), VSP International Science, 2000.

L. Devroye and A. Krzyżak, *New multivariate product density estimators*, 2000. Unpublished.

M. P. Gessaman, "A consistent nonparametric multivariate density estimator based on statistically equivalent blocks," *Annals of Mathematical Statistics*, vol. 41, pp. 1344–1346, 1970.

L. Gordon and R. Olshen, "Almost surely consistent nonparametric regression from recursive partitioning schemes," *Journal of Multivariate Analysis*, vol. 15, pp. 147–163, 1984.

G. Lugosi and A. Nobel, "Consistency of data-driven histogram methods for density estimation and classification," *Annals of Statistics*, vol. 24, pp. 687–706, 1996.

L. C. Zhao, P. R. Krishnaiah, and X. R. Chen, "Almost sure L_r-norm convergence for data-based histogram estimates," *Theory of Probability and its Applications*, vol. 35, pp. 396–403, 1990.

Chapter 13
Wavelet Estimates

§13.1. Definitions

In univariate curve estimation, wavelet systems have become popular. A wavelet system is a countable family of orthonormal functions (in $L_2(\mathbf{R})$) $\{\phi_k, \psi_{jk} : j \geq 0, k = 0, \pm 1, \pm 2, \ldots\}$ with the following properties:

(1) $\phi_k(x) = \phi(x-k)$, $k = 0, \pm 1, \pm 2, \ldots$, are translates of a fixed function ϕ in L_2 (called the father wavelet) such that the ϕ_k's are orthonormal.
(2) For $j \geq 0$, $\psi_{jk}(x) = 2^{j/2}\psi(2^j x - k)$, where ψ (the mother wavelet) is a function such that its translates $\psi(x-k)$, $k = 0, \pm 1, \pm 2, \ldots$, form an orthonormal basis of the space W_0 defined as the orthogonal complement of V_0 in V_1, where V_0 is the space generated by all translates of $\phi(x)$ and V_1 is the space generated by all translates of $\phi(2x)$.
(3) $\{\phi_k, \psi_{jk} : j \geq 0, k = 0, \pm 1, \pm 2, \ldots\}$ is an orthonormal basis of $L_2(\mathbf{R})$.

Then any $g \in L_2(\mathbf{R})$ may be written as

$$g(x) = \sum_{k=-\infty}^{\infty} \alpha_k \phi_k(x) + \sum_{j=0}^{\infty} \sum_{k=-\infty}^{\infty} \beta_{jk}\psi_{jk}(x),$$

where $\alpha_k(x) = \int g\phi_k$, $\beta_{jk} = \int g\psi_{jk}$, and the convergence is meant to be in $L_2(\mathbf{R})$. This expansion suggests the simplest wavelet density estimate

$$f_{n,t}(x) = \sum_{k=-\infty}^{\infty} \widehat{\alpha}_k \phi_k(x) + \sum_{j=0}^{t} \sum_{k=-\infty}^{\infty} \widehat{\beta}_{jk}\psi_{jk}(x),$$

where t is an arbitrary integer and the coefficients are calculated by

$$\widehat{\alpha}_k = \frac{1}{n}\sum_{i=1}^{n} \phi_k(x), \qquad \widehat{\beta}_{jk} = \frac{1}{n}\sum_{i=1}^{n} \psi_{jk}(x).$$

The simplest example of a wavelet system is the *Haar system* defined by

$$\phi(x) = \mathbf{1}_{[0,1]}(x) \text{ and } \psi(x) = \mathbf{1}_{[0,1/2]}(x) - \mathbf{1}_{(1/2,1]}(x).$$

The Haar wavelet density estimate has the advantage that $\int f_{n,t} = 1$. Recall from Corollary 10.1 that if the estimates do not integrate to one, the bounds we can guarantee are slightly worse. For the theory of constructing wavelet systems we refer to the texts by Daubechies (1992), Meyer (1993), and Härdle, Kerkyacharian, Picard, and Tsybakov (1998).

To understand the wavelet density estimate, consider the Haar system, and observe that the first term $\sum_{k=-\infty}^{\infty} \widehat{\alpha}_k \phi_k(x)$ is just a regular histogram estimate with cell width $h = 1$, while the terms $\sum_{k=-\infty}^{\infty} \widehat{\beta}_{jk} \psi_{jk}(x)$ are correction terms which account for changes in the density in intervals of length 2^{-t}. The fact that different translates of ψ are disjoint guarantees the local nature of the estimate. Interestingly, it is easy to see by induction on t that the estimate $f_{n,t}(x)$ is the same as the regular histogram estimate with bin width $h = 2^{-t}$ (Exercise 1). The number of terms t thus acts as a smoothing parameter. Also, by Exercise 10.2, the Haar wavelet estimate is universally consistent whenever $t \to \infty$ such that $2^t/n \to 0$. Wavelets have become popular because of their excellent approximation properties in Sobolev, Besov and other function spaces. In this chapter we study data-based selection of various parameters of wavelet density estimates, based on the methodology worked out in Chapter 10.

§13.2. Smoothing

The first obvious question concerns the selection of the parameter t for the estimate

$$f_{n,t}(x) = \sum_{k=-\infty}^{\infty} \widehat{\alpha}_k \phi_k(x) + \sum_{j=0}^{t} \sum_{k=-\infty}^{\infty} \widehat{\beta}_{jk} \psi_{jk}(x).$$

If both the father and mother wavelets are bounded, then $f_{n,t}(x)$ is a regular additive estimate, so we may use the estimate defined in Section 10.1 to select the parameter and t. To make the combinatorial argument simple, we limit the range of possible values of t by considering

$$\Theta = \{t : 1 \le t \le n\}.$$

For the Haar system this range clearly contains all meaningful values of t even for small sample sizes. The particular choice of the range $[1, n]$ is motivated by simplicity. Any polynomial function of n may replace n and lead to a similar result. Note that in the case of the Haar system, there is no need for restricting the value of t by Lemma 10.5. An easy counting argument shows that the number of Yatracos sets of the form

$$\{x : f_{n-m,\theta}(x) > f_{n-m,\theta'}(x) : \theta \ne \theta'\}$$

is not more than $(n+1)^2$, so that the shatter coefficient of the class $\mathcal{A}_\Theta$ of those sets is

$$S_{\mathcal{A}_\Theta}(m) \le (n+1)^2.$$

If the father and mother wavelets satisfy $\int \phi = 1$ and $\int \psi = 0$ (as is the case for the Haar wavelet estimate), then $\int f_{n,t} = 1$ for all t, and we may invoke Theorem 10.3 to conclude that the obtained density estimate f_n satisfies

$$\mathbf{E}\left\{\int |f_n - f|\right\} \leq 3 \inf_{\theta\in\Theta} \mathbf{E} \int |f_{n,\theta} - f| \left(1 + \frac{2m}{n-m} + 8\sqrt{\frac{m}{n}}\right) + 8\sqrt{\frac{2\log(n+1)+1}{m}} + \frac{3}{n}.$$

Thus, the selected estimate is close to optimal whenever the best choice for t provides a good estimate. In the case of the Haar basis we know that the class contains a universally consistent estimate, so that the estimate f_n is also universally consistent. If the father and mother wavelets fail to satisfy $\int \phi = 1$ and $\int \psi = 0$, then the factors of "3" should be replaced by "5".

For general wavelet systems we do not know whether there exists a choice of t such that $\mathbf{E} \int |f_{n,t} - f|$ converges to zero for all densities, even if one is allowed to choose t knowing the density f. In this respect, recall from Chapter 8 that other series estimates such as the trigonometric series estimate fail to be universally consistent for any choice of the number of terms of the estimate.

§13.3. Thresholding

The idea of wavelet thresholding is to smooth the simplest wavelet estimate defined in the previous section by canceling those terms from the wavelet expansion whose estimated coefficients are small. More precisely, consider the infinite family of estimates

$$f_{n,\tau}(x) = \sum_{k=-\infty}^{\infty} \widehat{\alpha}_k \phi_k(x) + \sum_{j=0}^{t} \sum_{k=-\infty}^{\infty} F_\tau\left(\widehat{\beta}_{jk}\right)\psi_{jk}(x),$$

where now t is fixed for simplicity, the threshold parameter τ is any positive number, and

$$F_\tau(z) = z\mathbf{1}_{[|z|>\tau]}, \quad z \in \mathbf{R}.$$

In other words, all coefficients $\widehat{\beta}_{jk}$ whose absolute value falls below the threshold $\tau > 0$ are set to zero. We call these estimates *thresholding wavelet estimates*. If the estimate is based on the Haar basis, then $f_{n,0}$ (no thresholding) is just the regular histogram estimate. In this case thresholding may be thought of as a local smoothing procedure which unites certain neighboring cells of the histogram if their cardinalities are close to each other.

Our goal is to devise an optimal data-dependent way of selecting the threshold τ. Once again, the minimum distance estimate of Section 10.1 can be applied automatically, and the only work left to do is to make sure that the complexity of the corresponding Yatracos class is not too large. This is indeed the case for the Haar system, as the next result shows. The selection of τ for more general wavelet systems, as well as the joint selection of τ and the parameter t, are treated in the exercises.

THEOREM 13.1. *Consider the class of thresholding wavelet estimates $\{f_{n,\tau} : \tau > 0\}$ based on the Haar system (i.e., $\phi(x) = \mathbf{1}_{[0,1]}(x)$ and $\psi(x) = \mathbf{1}_{[0,1/2]}(x) - \mathbf{1}_{(1/2,1]}(x)$) with all possible positive values of τ but with t fixed. The shatter coefficient of the Yatracos class $\mathcal{A}$ of the sets*

$$A_{\tau,\tau'} = \{x : f_{n-m,\tau}(x) > f_{n-m,\tau'}(x) : \tau \neq \tau'\}$$

is bounded by

$$\mathrm{S}_{\mathcal{A}}(m) \leq m^2(t+1)^2,$$

and consequently the minimum distance density estimate f_n satisfies, for all densities f on **R**, *and for all $m \leq n/2$:*

$$\mathbf{E}\left\{\int |f_n - f|\right\} \leq 3 \inf_{\tau>0} \mathbf{E} \int |f_{n,\tau} - f| \left(1 + \frac{2m}{n-m} + 8\sqrt{\frac{m}{n}}\right) + 8\sqrt{\frac{2\log(m(t+1)) + 1}{m}} + \frac{3}{n}.$$

PROOF. First note that for each $\tau > 0$, $f_{n,\tau}$ is a regular additive estimate, so Theorem 10.3 is applicable, and the only thing required is the bound for the shatter coefficient. Fix $y_1, \ldots, y_m \in \mathbf{R}$. As we increase τ from zero to infinity, the value of the vector

$$z_\tau = \left(f_{n-m,\tau}(y_1), \ldots, f_{n-m,\tau}(y_m)\right)$$

can take at most $(t+1)m$ different values, since the value of each $f_{n-m,\tau}(y_m)$ is influenced by at most $t+1$ coefficients $\widehat{\beta}_{jk}$, and each $\widehat{\beta}_{jk}$ is set to zero at one value of τ and it never changes back again as τ is increased. Denote the values of τ at which z_τ changes by $\tau^{(1)} < \tau^{(2)} < \cdots < \tau^{(M)}$, where $M \leq tm$. These values determine a partition of **R** into $M+1 \leq (t+1)m$ intervals. The Cartesian product of this partition with itself determines a partition of $[0,\infty)^2$ into $(M+1)^2$ rectangles. If the pairs (τ, τ') and (τ_1, τ_1') fall in the same rectangle, then

$$\mathbf{1}_{y_i \in A_{\tau,\tau'}} = \mathbf{1}_{y_i \in A_{\tau_1,\tau_1'}} \quad \text{for all } i = 1, \ldots, m.$$

It follows that the shatter coefficient $\mathrm{S}_{\mathcal{A}}(m)$ is bounded by $m^2(t+1)^2$. □

§13.4. Soft Thresholding

A smoother variant of the thresholding wavelet estimate of the previous section is obtained if we modify $F_\tau(z) = z\mathbf{1}_{[|z|>\tau]}$ to

$$F_\tau(z) = (z+\tau)\mathbf{1}_{[z>\tau]} + (z-\tau)\mathbf{1}_{[z<-\tau]}, z \in \mathbf{R},$$

where τ is a positive parameter as before. Soft thresholding prevents abrupt changes in the density estimate $f_{n,\tau}$ if the parameter τ is changed by a small amount. $f_{n,\tau}$ is called a *soft thresholding wavelet estimate.* Various data-independent and data-dependent ways of selecting τ have been proposed in the literature. The purpose of this section is to show that, once again, the minimum distance estimate of Chapter 10 is capable of selecting a close-to-optimal value for τ for the soft thresholding wavelet estimate. Recall once more that in order to obtain a performance bound as in Theorem 2, it suffices to derive a polynomial upper bound for the shatter coefficients of the corresponding Yatracos class. This is indeed possible, as is shown in Theorem 2 below. For concreteness, we again focus on the Haar system. The extension to compactly supported father and mother wavelets is straightforward.

THEOREM 13.2. *Consider the class of soft thresholding wavelet estimates $\{f_{n,\tau} : \tau > 0\}$ based on the Haar system (i.e., $\phi(x) = \mathbf{1}_{[0,1]}(x)$ and $\psi(x) = \mathbf{1}_{[0,1/2]}(x) - \mathbf{1}_{(1/2,1]}(x)$) with all possible positive values of τ but with t fixed. The shatter coefficient of the Yatracos class $\mathcal{A}$ of the sets*

$$A_{\tau,\tau'} = \{x : f_{n-m,\tau}(x) > f_{n-m,\tau'}(x) : \tau \neq \tau'\}$$

is bounded by

$$\mathbb{S}_{\mathcal{A}}(m) \le 2(m+1)^5(t+1)^2.$$

PROOF. Fix $y_1, \ldots, y_m \in \mathbf{R}$, and note that for each y_i, the value of $f_{n-m,\tau}(y_i)$ is influenced by at most $t+1$ of the $\widehat{\beta}_{jk}$'s. Consider the set S of those values of $|\widehat{\beta}_{jk}|$'s which can influence the value of at least one $f_{n-m,\tau}(y_i)$. The cardinality of this set is at most $m(t+1)$. Consider the set of pairs (τ,τ') such that $0 < \tau < \tau'$. The numbers in S are partitioned into three intervals $[0,\tau)$, $[\tau,\tau')$, $[\tau',\infty)$. The number of different ways this can be done for different choices of the pair (τ,τ') is at most $\binom{|S|+2}{2} \le m^2(t+1)^2$ (if $m > 3$). Thus, we obtain a partition of the set $[0,\infty)^2$ of all possible pairs (τ,τ') into not more than $2m^2(t+1)^2$ cells such that in each cell all pairs induce the same partition of the numbers in S. The factor 2 comes from now considering all pairs $\tau \neq \tau'$, and not only those with $\tau < \tau'$. Consider now one such cell in $[0,\infty)^2$. We may assume without loss of generality that $\tau < \tau'$. Writing

$$
\begin{aligned}
&f_{n-m,\tau}(y_i) - f_{n-m,\tau'}(y_i) \\
&\quad = \sum_{k,j:\widehat{\beta}_{jk}>\tau'} (\tau' - \tau)\psi_{jk}(y_i) + \sum_{k,j:\widehat{\beta}_{jk}\in(\tau,\tau']} (\widehat{\beta}_{jk} - \tau)\psi_{jk}(y_i) \\
&\quad + \sum_{k,j:\widehat{\beta}_{jk}<-\tau'} (\tau - \tau')\psi_{jk}(y_i) + \sum_{k,j:\widehat{\beta}_{jk}\in[-\tau',\tau)} (\widehat{\beta}_{jk} + \tau)\psi_{jk}(y_i),
\end{aligned}
$$

we see that within one cell, this has the form

$$
f_{n-m,\tau}(y_i) - f_{n-m,\tau'}(y_i) = A(y_i)\tau + B(y_i)\tau' + C(y_i),
$$

where the coefficients $A(y_i), B(y_i), C(y_i)$ do not depend on (τ, τ'). Therefore, as (τ, τ') is varied within one cell, the vector

$$
\left(\mathbf{1}_{\left[f_{n-m,\tau}(y_1)-f_{n-m,\tau'}(y_1)>0\right]}, \ldots, \mathbf{1}_{\left[f_{n-m,\tau}(y_m)-f_{n-m,\tau'}(y_m)>0\right]}\right)
$$

can take at most as many values as the number of different ways m points in $\mathbf{R}^3$ (with components $(A(y_i), B(y_i), C(y_i))$ can be dichotomized by planes containing the origin. By Lemma 4.2 this is not more than $(m+1)^3$. Considering all $2m^2(t+1)^2$ cells, the total number of different values these vectors can take is not more than $2(m+1)^3 m^2 (t+1)^2$. $\square$

Several other alternatives for wavelet thresholding have been proposed in the literature. Most of them can be treated by similar arguments as in Theorems 1 and 2 above. An example is given in the exercises.

§13.5. Bibliographic Remarks

The general theory of wavelets is treated in texts by Daubechies (1992), Meyer (1993), and Igari (1998). The use of wavelets in various curve estimation problems is surveyed in Härdle, Kerkyacharian, Picard, and Tsybakov (1998), where approximation properties of wavelets are discussed in detail. Wavelet thresholding was pioneered by Donoho, Johnstone, Kerkyacharian, and Picard (1995, 1996). We refer to Härdle, Kerkyacharian, Picard, and Tsybakov (1998) for a survey.

§13.6. Exercises

Exercise 13.1 Prove that the Haar wavelet estimate $f_{n,t}$ is just a regular histogram estimate with bin width $h = 2^{-t}$.

Exercise 13.2 Consider the family of thresholding wavelet estimates of Section 2 based on a father wavelet and a mother wavelet which are bounded and supported in $[0,1]$, but $\int \phi = 1$ and $\int \psi = 0$ are not necessarily satisfied. Thus, in order to make them eligible for the application of Theorem 10.3, the estimates should be normalized so that their integral becomes one, that is,

$$f_{n,\tau}(x) = c_\tau \left(\sum_{k=-\infty}^{\infty} \widehat{\alpha}_k \phi_k(x) + \sum_{j=0}^{t} \sum_{k=-\infty}^{\infty} F_\tau \left(\widehat{\beta}_{jk} \right) \psi_{jk}(x) \right),$$

$\int f_{n,\tau} = 1$. Show that the estimate $\mathbf{S}_{\mathcal{A}}(m) \leq m^2(t+1)^2$ for the shatter coefficient of the Yatracos class remains valid.

Exercise 13.3 How can the bound of Theorem 1 be modified if instead of the Haar system, we use wavelets based on a mother wavelet supported in an interval of length N?

Exercise 13.4 Work out an adequate estimate for the joint selection of the parameter $1 \leq t \leq n$ and the threshold $\tau > 0$ for the thresholding wavelet estimate.

Exercise 13.5 (GLOBAL THRESHOLDING). Consider the globally thresholding wavelet estimate defined by

$$f_{n,\tau}(x) = \sum_{k=-\infty}^{\infty} \widehat{\alpha}_k \phi_k(x) + \sum_{j=0}^{t} F_\tau \left(\sum_{k=-\infty}^{\infty} \widehat{\beta}_{jk} \psi_{jk}(x) \right),$$

where t is fixed, the threshold parameter τ is any positive number, and

$$F_\tau(z) = z \mathbf{1}_{[|z|>\tau]}, z \in \mathbf{R}.$$

Assume that the estimate is based on the Haar system. Obtain a polynomial upper bound for the shatter coefficient $\mathbf{S}_{\mathcal{A}}(m)$ of the class $\mathcal{A}$ of sets of the form

$$A_{\tau,\tau'} = \{x : f_{n-m,\tau}(x) > f_{n-m,\tau'}(x) : \tau \neq \tau'\}.$$

§13.7. References

I. Daubechies, *Ten Lectures on Wavelets*, SIAM, Philadelphia, PA, 1992.

D. Donoho, I. Johnstone, G. Kerkyacharian, and D. Picard, "Wavelet shrinkage: Asymptotia?," *Journal of the Royal Statistical Society, Series B*, vol. 57, pp. 301–369, 1995.

D. Donoho, I. Johnstone, G. Kerkyacharian, and D. Picard, "Density estimation by wavelet thresholding," *Annals of Statistics*, vol. 24, pp. 508–539, 1996.

W. Härdle, G. Kerkyacharian, D. Picard, and A. Tsybakov, *Wavelets, Approximation, and Statistical Applications*, Springer-Verlag, New York, 1998.

S. Igari, *Real Analysis—With an Introduction to Wavelet Theory*, American Mathematical Society, Providence, RI, 1998.

Y. Meyer, *Wavelets: Algorithms and Applications*, SIAM, Philadelphia, PA, 1993.

Chapter 14
The Transformed Kernel Estimate

§14.1. The Transformed Kernel Estimate

The transformed kernel estimate on the real line was introduced in an attempt to reduce the L_1 error in a relatively cheap manner. The data are first transformed $T : \mathbf{R} \to \mathbf{R}$ by a strictly monotonically increasing almost everywhere differentiable transformation T: $Y_1 = T(X_1), \ldots, Y_n = T(X_n)$. The density of Y_1 is

$$g(y) = f(T^{-1}(y))\, T^{-1\prime}(y),$$

where T^{-1} denotes the inverse of T. g is estimated by the kernel estimate

$$g_n(y) = \frac{1}{nh} \sum_{i=1}^{n} K\left(\frac{y - Y_i}{h}\right),$$

where K is a kernel with $\int K = 1$. Finally, the estimate is subjected to the inverse transformation, thus obtaining the transformed kernel estimate

$$f_{n,T}(x) = g_n(T(x))T'(x) = \frac{1}{nh} \sum_{i=1}^{n} K\left(\frac{T(x) - T(X_i)}{h}\right) T'(x).$$

Clearly, $\int f_{n,T} = 1$. As this leaves the L_1 error unaltered, that is,

$$\int |f_{n,T}(x) - f(x)|\, dx = \int |g_n(y) - g(y)|\, dy,$$

it suffices to study the L_1 error in the transformed space, and hence the interest of such estimates. If $T = ax + b$ is linear, then $f_{n,T}$ is just the ordinary kernel estimate with smoothing factor $h = 1/a$. The interest, therefore, is in nonlinear transformations. It is known that heavy tails are to be avoided for kernel estimates (Devroye and Györfi, 1985). Thus, we need transformations that compactify the data, where the transformation is selected by the data. If we do not put any restriction on the form of the transformation, optimal choice becomes impossible (see Exercise 1).

In this chapter we study the automatic choice of the transformation from certain classes of transformations. Our main example is the Box–Cox family of transformations with which statisticians and data analysts are familiar. We will show that we can jointly pick the transformation and the smoothing factor h in an asymptotically optimal manner, still modulo a factor 3, without placing any restrictions on the density or the parameters. The transformations considered here are useful to treat tail problems. In our second example, the transformation is piecewise linear. Such estimators are close in spirit to the variable kernel estimators discussed in Chapter 12.

§14.2. Box–Cox Transformations

Consider the family $\{T_a : a \in [0,1]\}$ of transformations defined, for $x > 0$, by

$$T_a(x) = \begin{cases} \frac{x^a - 1}{a} & \text{if } a > 0 \\ \log x & \text{if } a = 0. \end{cases}$$

These functions are often used to transform the (nonnegative) data so that large tails become manageable. We consider the kernel estimate for the transformed data. In particular, we study the joint data-based selection of the tranformation (i.e., the value of a) and the bandwidth h. For simplicity, we only consider the naive kernel $K = \mathbf{1}_{[-1/2,1/2]}$. Therefore, the class of estimates $\{f_{n,\theta} : \theta \in \Theta\}$ is defined by

$$f_{n,\theta}(x) = \frac{1}{nh} \sum_{i=1}^{n} \mathbf{1}_{[|T_a(x) - T_a(X_i)| \le h/2]} x^{a-1},$$

where $\theta = (a,h)$ and $\Theta = [0,1] \times (0,\infty)$. Note that we assume that all data points are positive and that $f_{n,\theta}(x)$ is only defined for $x > 0$. Again, to see if the proposed parameter selection method works, it suffices to bound $\mathrm{S}_{\mathcal{A}_\Theta}(m)$.

LEMMA 14.1. *Let $\mathcal{A}_\Theta$ denote the Yatracos class corresponding to the family of kernel estimates on $\mathbf{R}^+$ based on all Box–Cox transformations T_a, $a \in [0,1]$ and all smoothing factors $h > 0$. If $m \ge 2$ and $n - m \ge 2$, then*

$$\mathrm{S}_{\mathcal{A}_\Theta}(m) \le 9/4\, m^6 (n-m)^4.$$

LEMMA 14.2. *If $b_1, \ldots, b_k, c_1, \ldots, c_k \in \mathbf{R}$, then the function*

$$g(x) = \sum_{i=1}^{k} b_i e^{c_i x}$$

is either identically zero or takes the value 0 at strictly fewer than k places.

PROOF. Denote $y = e^x$, so that we have to bound the number of roots of $\tilde{g}(y) = \sum_{i=1}^k b_i y^{c_i}$ for $y \in (0, \infty)$. Without loss of generality we may assume that $b_k \neq 0$. Note first that $\tilde{g}$ cannot be identically zero on any interval of nonzero length. Let $Z(f)$ denote the number of positive roots of a function f. Then we have

$$
\begin{aligned}
Z\left(\sum_{i=1}^k b_i y^{c_i}\right) &= Z\left(\sum_{i=1}^k b_i y^{a_i}\right) \qquad \text{(where } a_i = c_i - c_1, \text{ all } i; \text{ thus, } a_1 = 0)\\
&\leq Z\left(\sum_{i=2}^k b_i a_i y^{a_i - 1}\right) + 1\\
&\qquad \text{(if } g \text{ is continuously differentiable then } Z(g) \leq 1 + Z(g'))\\
&= Z\left(\sum_{i=2}^k b_i' y^{c_i'}\right) + 1\\
&\qquad \text{(where } b_i' = b_i a_i, c_i' = a_i - a_2,\ i \geq 2).
\end{aligned}
$$

Note that the c_i' are increasing, $c_2' = 0$, and $b_i' \neq 0$. As $Z\left(b_k y^{c_k}\right) = 0$ for $b_k \neq 0$, we derive our claim by induction on k. □

PROOF OF LEMMA 1. Consider the $m \times (n - m)$ matrix $z_{a,h}$ with entries

$$z_{a,h}^{t,i} = \mathbf{1}_{[|X_i^a - y_t^a| < ah]}, \quad t = 1, \ldots, m, \quad i = 1, \ldots, n - m.$$

First we bound the number of possible different values of the matrix $z_{a,h}$ as $(a, h) \in [0, 1] \times (0, \infty)$. Observe that in the set $[0, 1] \times (0, \infty)$, each pair (t, i) defines a curve given by

$$u^a - v^a - ah = 0 \qquad \text{where } u = \max(X_i, y_t) \text{ and } v = \min(X_i, y_t).$$

If two curves $u^a - v^a - ah = 0$ and $w^a - z^a - ah = 0$ intersect at the point (a, h), then

$$e^{a \log u} - e^{a \log v} - e^{a \log w} + e^{a \log z} = 0.$$

According to Lemma 2, this cannot happen at more than three points (unless $u = w$ and $v = z$). Next we argue that these curves partition the set $[0, 1] \times (0, \infty)$ into at most $(3m^2(n - m)^2 - m(n - m) + 4)/2 \leq 3/2 m^2(n - m)^2$ connected regions. This may be seen by induction: if S_N denotes the number of connected regions defined by N such curves, then it is clear that $S_1 = 2$ and $S_{N+1} \leq S_N + 3N + 1$, since any two curves intersect at at most three points. The solution of this recursion is $S_N \leq (3N^2 - N + 2)/2$. Since inside each region the value of the matrix $z_{a,h}$ is a constant, $3/2 m^2(n - m)^2$ is an upper bound on the number of possible values of the matrix. Therefore,

$$|\{(z_{a,h}, z_{a',h'}) : a, a' \in [0, 1], h, h' > 0\}| \leq 9/4 m^4(n - m)^4.$$

Consider now a region in $([0,1] \times (0,\infty))^2$ over which $(z_{a,h}, z_{a',h'})$ is constant, with value, say, (w, w'), and denote the set of such quadruples (a, h, a', h') by $U_{(w,w')}$. Denoting $\theta = (a, h)$ and $\theta' = (a', h')$:

$$y_i \in A_{\theta,\theta'} \text{ if and only if } \frac{y_t^{a-1}}{h} \sum_{i=1}^{n-m} w^{(t,i)} \geq \frac{y_t^{a'-1}}{h'} \sum_{i=1}^{n-m} w'^{(t,i)},$$

or, equivalently, if and only if

$$(a-1)\log y_t - \log h + \log\left(\sum_{i=1}^{n-m} w^{(t,i)}\right)$$
$$\geq (a'-1)\log y_t - \log h' + \log\left(\sum_{i=1}^{n-m} w'^{(t,i)}\right).$$

Therefore, denoting $W_t = \sum_{i=1}^{n-m} w^{(t,i)}$ and $W'_t = \sum_{i=1}^{n-m} w'^{(t,i)}$, the maximal number of different values of

$$(\mathbf{1}_{A_{\theta_1,\theta_2}}(y_1), \ldots, \mathbf{1}_{A_{\theta_1,\theta_2}}(y_m))$$

is at most the number of different values of the vector

$$\left(\mathbf{1}_{[(a-a')\log y_1 - \log(h/h') + \log(W_1/W'_1) \geq 0]}, \ldots,\right.$$
$$\left.\mathbf{1}_{[(a-a')\log y_m - \log(h/h') + \log(W_m/W'_m) \geq 0]}\right)$$

takes as a, a', h, h' all vary through $\mathbf{R}^+$. But this is not more than the maximal number of different ways of dichotomizing m points by two-dimensional hyperplanes, which, by Corollary 4.2, is at most m^2 (since $m \geq 2$). Conclude by collecting bounds. □

Lemma 1 and Theorem 10.3 yield the following bound:

THEOREM 14.1. *Assume that the estimate of Section 10.1 is used to simultaneously select the Box–Cox transformation T_a, $a \in [0,1]$, and the smoothing factor $h > 0$ for the transformed kernel estimate*

$$f_{n,a,h}(x) = \frac{1}{nh} \sum_{i=1}^{n} \mathbf{1}_{[|T_a(x) - T_a(X_i)| \leq h/2]} x^{a-1}.$$

If f_n denotes the obtained density estimate, then for all $n, m \leq n/2$, and each density f over $\mathbf{R}^+$:

$$\mathbf{E}\left\{\int |f_n - f|\right\}$$
$$\leq 3 \inf_{a\in[0,1],h>0} \mathbf{E}\left\{\int |f_{n,a,h} - f|\right\}\left(1 + \frac{2m}{n-m} + 8\sqrt{\frac{m}{n}}\right)$$
$$+ \sqrt{\frac{8\log((9/2)m^6(n-m)^4)}{m}}.$$

The estimate of Theorem 1 is asymptotically at least as good as the best untransformed density estimate $f_{n,h}$ (modulo a factor of three) for all densities. In other words,

$$\sup_f \limsup_{n\to\infty} \frac{\mathbf{E}\{\int |f_n - f|\}}{\inf_{h>0} \mathbf{E}\{\int |f_{n,h} - f|} \} \leq 3.$$

On the other hand, for a large family of heavy-tailed densities,

$$\limsup_{n\to\infty} \frac{\mathbf{E}\left\{\int |f_n - f|\right\}}{\inf_{h>0} \mathbf{E}\left\{\int |f_{n,h} - f|\right\}} = 0.$$

To see this, just recall from Section 9.7 that for all densities on $[0,\infty)$, whose tail behaves like $f(x) \sim x^{-b}$ for some $b < 2$, we have

$$n^{2/5} \inf_{h>0} \mathbf{E}\left\{\int |f_{n,h} - f|\right\} \to \infty.$$

On the other hand, whenever $a < (b-1)/2$, the transformed density

$$g(y) = f(T_a^{-1}(y))\, T_a^{-1\prime}(y)$$

satisfies the tail condition $\int \sqrt{g} < \infty$, which is necessary to achieve the optimal $n^{-2/5}$ rate of convergence. Since $\int |f_{n,T}(x) - f(x)|\, dx = \int |g_n(y) - g(y)|\, dy$, the transformed estimate indeed achieves the indicated asymptotic superiority with respect to the kernel estimate whenever $\int |g''| < \infty$.

§14.3. Piecewise Linear Transformations

In an alternative example, we assume that each T in the collection is piecewise linear with at most r pieces. More precisely, T has the form

$$T(x) = \sum_{j=1}^{r} (a_j x + b_j) \mathbf{1}_{B_j}(x),$$

where $B_1, \ldots, B_r$ is a partition of $\mathbf{R}$ into r intervals defined by the endpoints $c_1 \leq \cdots \leq c_{r-1} \in \mathbf{R}$, and $a_1, \ldots, a_r > 0$, $b_1, \ldots, b_r \in \mathbf{R}$ are real constants. Of course, we have to restrict the parameters such that T is continuous and strictly increasing. Assume for simplicity that $K = \mathbf{1}_{[-1/2,1/2]}$. The next combinatorial calculation gives us the necessary tool for deriving a performance bound for the transformed kernel estimate with piecewise linear transformation, if the transformation is selected by the minimum distance method of Section 10.1.

LEMMA 14.3. *If Θ denotes the set of all possible $(3r-1)$-dimensional vectors $(a_1,\ldots,a_r,b_1,\ldots,b_r,c_1,\ldots,c_{r-1})$ and $m(n-m)\geq 2$, then*

$$\mathbb{S}_{\mathcal{A}_\Theta}(m) \leq (m+1)^{2r+1}\left[\binom{n-1}{r-1}((n-m)m)^{4r^2}\right]^2 \leq (m+1)^{8r^2+4r+1}n^{8r^2+2r}.$$

PROOF. For each $\theta\in\Theta$, we define an $m\times r$ matrix z_θ with entries

$$z_\theta^{(t,j)} = 1_{B_j}(y_t)\sum_{i=1}^{n-m} 1_{[-1/2,1/2]}\left(a_j y_t + b_j - \sum_{j'=1}^{r}(a_{j'}X_i + b_{j'})1_{B_{j'}}(X_i)\right),$$

where $t\leq m$ and $j\leq r$. First we bound the number of different values the matrix z_θ can take as θ ranges through Θ. First consider the $r(m+(n-m))$-vector of the indicators $1_{B_j}(y_t), 1_{B_j}(X_i)$, $t\leq m$, $i\leq n-m$, $j\leq r$. Clearly, the number of different values this vector can take is not more than the number of different ways $m+(n-m)$ points can be partitioned into r intervals. This is at most $\binom{n-1}{r-1}$. Consider now a set of partitions into r intervals such that the indicators $1_{B_j}(y_t)$ and $1_{B_j}(X_i)$ are fixed, $t\leq m, i\leq n-m, j\leq r$. By the above argument there are at most $\binom{n-1}{r-1}$ such sets. Over such a set, the matrix z_θ cannot take more distinct values than the $(n-m)\times m$ matrix with entries

$$1_{[-1/2,1/2]}(a_{j_t}y_t + b_{j_t} - a_{j(i)}X_i - b_{j(i)}),\ i\leq n-m,\ t\leq m,$$

where j_t is the index of the set B_j containing y_t, and $j(i)$ is the index of the set B_j containing X_i. But since the number of different values of the pair $(j_t, j(i))$ is at most r^2, the number of values the above matrix can take does not exceed the number of different ways $(n-m)m$ two-dimensional points (formed by the pairs (X_i,y_t)) can be dichotomized by $2r^2$ hyperplanes, which, by Corollary 4.2 is at most $((n-m)m)^{4r^2}$. Therefore, the number of different values of the matrix z_θ is not more than $\binom{n-1}{r-1}((n-m)m)^{4r^2}$.

From this point on we proceed as in previous proofs:

$$|\{(z_{\theta_1},z_{\theta_2}) : \theta_1,\theta_2\in\Theta\}| \leq \left[\binom{n-1}{r-1}((n-m)m)^{4r^2}\right]^2.$$

Let $\mathcal{W} = \{(w,w') : (w,w') = (z_{\theta_1},z_{\theta_2})$ for some $\theta_1,\theta_2\in\Theta\}$. For fixed $(w,w')\in\mathcal{W}$, let $U_{(w,w')}$ denote the collection of all (θ_1,θ_2) such that $(z_{\theta_1},z_{\theta_2}) = (w,w')$. For $(\theta_1,\theta_2)\in U_{(w,w')}$, we have

$$y_t\in A_{\theta_1,\theta_2}\quad\text{if and only if}\quad \sum_{j=1}^{r} a_{j,1}z_{\theta_1}^{(t,j)} \geq \sum_{j=1}^{r} a_{j,2}z_{\theta_2}^{(t,j)},$$

where $\theta_i = (a_{1,i}, \ldots, a_{r,i}, b_{1,i}, \ldots, b_{r,i}, c_{1,i}, \ldots, c_{r-1,i})$ for $i = 1, 2$. Within the set $U_{(w,w')}$, $z_{\theta_1}^{(t,j)}$ and $z_{\theta_2}^{(t,j)}$ are fixed for all t and j, and therefore the number of different values of the binary vector

$$(\mathbf{1}_{A_{\theta_1,\theta_2}}(y_1), \ldots, \mathbf{1}_{A_{\theta_1,\theta_2}}(y_m))$$

is at most the number of different values of vector

$$\left(\mathbf{1}_{\sum_{j=1}^{r}\left(a_{j,1}z_{\theta_1}^{(1,j)} - a_{j,2}z_{\theta_2}^{(1,j)}\right)\geq 0}, \ldots, \mathbf{1}_{\sum_{j=1}^{r}\left(a_{j,1}z_{\theta_1}^{(m,j)} - a_{j,2}z_{\theta_2}^{(m,j)}\right)\geq 0}\right)$$

takes as $a_{1,1}, \ldots, a_{r,1}, a_{1,2}, \ldots, a_{r,2}$ all vary through $\mathbf{R}^+$. But this is not more than the maximal number of different ways of dichotomizing m points in $\mathbf{R}^{2r}$ by hyperplanes, which, according to Corollary 4.2, is at most $(m+1)^{2r+1}$. The proof of is finished. □

§14.4. Bibliographic Remarks

The transformed kernel estimate was introduced by Devroye and Györfi (1985). Tranforming the kernel estimate using the Box–Cox family of transformations was suggested by Wand, Marron and Ruppert (1991) and Wand and Jones (1995). For practical data-based versions of other transformations, we refer to Wand, Marron and Ruppert (1991) and Ruppert and Cline (1994). Theorem 1 appears also in Devroye, Lugosi, and Udina (2000), who report experimental results on the estimate.

§14.5. Exercises

Exercise 14.1 Show that for any data-based choice T_n of the transformation T:

$$\sup_f \frac{\mathbf{E}\{\int |f_{n,T_n}(x) - f(x)|\,dx\}}{\inf_T \mathbf{E}\{\int |f_{n,T}(x) - f(x)|\,dx\}} \to \infty,$$

where the infimum is taken over all strictly monotone differentiable transformations $T : \mathbf{R} \to \mathbf{R}$. HINT: Proceed as in Theorem 15.3.

Exercise 14.2 Consider the one-parameter family of transformations

$$T_b(x) = \sqrt{|x-b|}\ \mathrm{sgn}(x-b),\ b \in \mathbf{R}.$$

Derive a bound similar to that of Lemma 1 for the joint selection of the parameter b and the bandwidth h for the transformed kernel estimate. Argue that the obtained estimate improves on the rate of convergence of the ordinary kernel estimate for certain densities with one thick infinite peak.

Exercise 14.3 Consider the two-dimensional Box–Cox transformation of $\mathbf{R}^+ \times \mathbf{R}^+$ defined by $T(x,y) = (T_a(x), T_b(y))$, where T_a and T_b

are one-dimensional Box–Cox transformations. Start with a kernel estimate with window $K(x,y) = \mathbf{1}_{[-1/2,1/2]}(x)\mathbf{1}_{[-1/2,1/2]}(y)$, and bandwidths h_1 and h_2. Obtain extensions of Lemma 1 first for the parameter space given by $a, b \in [0,1]; h_1 = h_2 > 0$, and then for the parameter space defined by $a, b \in [0,1]; h_1, h_2 > 0$. Reformulate the estimate and bound of Theorem 1.

§14.6. References

L. Devroye and L. Györfi, *Nonparametric Density Estimation: The L_1 View*, Wiley, New York, 1985.

L. Devroye, L. Györfi, and G. Lugosi, *A Probabilistic Theory of Pattern Recognition*, Springer-Verlag, New York, 1996.

L. Devroye, G. Lugosi, and F. Udina, "Inequalities for a new data-based method for selecting nonparametric density estimates," in: *Festschrift in Honour of George Roussas* (edited by M. L. Puri), VSP International Science, 2000.

D. Ruppert and D. B. H. Cline, "Bias reduction in kernel density estimation by smoothed empirical transformations," *Annals of Statistics*, vol. 22, pp. 185–210, 1994.

M. P. Wand and M. C. Jones, *Kernel Smoothing*, Chapman & Hall, London, 1995.

M. P. Wand, J. S. Marron, and D. Ruppert, "Transformations in density estimation," *Journal of the American Statistical Association*, vol. 86, pp. 343–361, 1991.

Chapter 15
Minimax Theory

§15.1. Estimating a Density from One Data Point

Let us use an extreme example to start off this chapter. Assume that we are shown one data point X drawn from an unknown density f that belongs to a class $\mathcal{F}$, which is given to us. Our estimate of f is $g(x, X)$, where g should be such that it is best possible. As we do not know f, we naturally assume that after g is picked, our adversary picks the worst $f \in \mathcal{F}$. Thus, the problem is to find a mapping g such that

$$\sup_{f\in\mathcal{F}} \int f(y) \left(\int |g(x,y) - f(x)|\, dx \right) dy$$

is minimal. For a particular g, the above quantity is called the worst-case risk for g, denoted by $R(g, \mathcal{F})$. This is the minimax point of view: we try to find the best solution taking a pessimistic view of the estimated density. In general, the exact solution is very hard to find. In view of our remarks in Section 5.6 about normalization, it is clear in the first place that the optimal g always is a density in x, for every y. The quantity $R(\mathcal{F}) = \inf_g R(g)$ is called the minimax risk, and depends only on $\mathcal{F}$.

If $|\mathcal{F}| = 1$, then we obviously take $g(x, y) = f(x)$, and disregard even the one data point shown us. If $\mathcal{F} = \{f_1, f_2\}$, then there are several possibilities: without even looking at X, one might consider the *average* $g(x, y) = (f_1(x) + f_2(x))/2$. In that case, $R(g, \mathcal{F}) = J/2$, where $J = \int |f_1 - f_2|$. This is the standard to compare against. In fact, if the sample size $n = 0$, it is the best we can do. Next, consider a choice $g \in \mathcal{F}$. The *decisive estimate* is based on a decision which of f_1 or f_2 to pick, based on the sample X. There exists a set A such that $g(x, y) = f_1(x)1_A(y) + f_2(x)1_{A^c}(y)$. We have

$$R(g, \mathcal{F}) = \max\left\{ \int_A f_2, \int_{A^c} f_1 \right\} \times \int |f_1 - f_2|$$

and indeed, this is minimal when $A = \{f_1 > f_2\}$, our good old Scheffé set. Recall that f_1 and f_2 are given, so A is known. So, if we set $J = \int |f_1 - f_2| = 2\int_A (f_1 - f_2)$, then we have for this particular choice of A,

$$\frac{J(2-J)}{4} = \frac{J\left(1+\int_A(f_2-f_1)\right)}{2} \le J\max\left(\int_A f_2, \int_{A^c} f_1\right)$$
$$= R(g,\mathcal{F}) \le J\left(1+\int_A (f_2-f_1)\right) = \frac{J(2-J)}{2},$$

where we used the fact that the maximum exceeds the average. The above inequalities imply that $R(\mathcal{F}) \le J(2-J)/2$, because we exhibited one prototype function g. In fact, the method of constructing prototype estimates to get upper bounds on the minimax risk is virtually the only tool available to us. Unfortunately, we do not have a lower bound for $R(\mathcal{F})$ because we failed to show that the optimal g is indeed of the form shown above: $g \in \mathcal{F}$. And in fact, we were blatantly wrong in our hypothesis! In the exercises, a third estimate dubbed the *hesitator* is considered: it forms a convex mixture between f_1 and f_2 with the mixture weight decided by X. Consider the following particular example in which the best estimator is definitely not in $\mathcal{F}$: let f_1 and f_2 be uniform on $[0,1]$ and $[1/2,1]$, respectively. Note that $J = \int |f_1 - f_2| = 1$. The Scheffé set $A = \{f_1 > f_2\}$ is $[0, 1/2]$, and the best worst-case risk with $g \in \mathcal{F}$ is $J\max\left(\int_A f_2, \int_{A^c} f_1\right) = 1/2$. This is not better than the average estimate which ignores the data! Now consider the following competing density estimate: if $X \in [0, 1/2]$, we set $g = f_1$. If $X \in (1/2, 1]$, we set $g = 5/3\mathbf{1}_{[1/2,1]}(x) + 1/3\mathbf{1}_{[0,1/2]}(x)$. Verify that for both densities f_1 and f_2, the expected error is $1/3$ and thus, the worst-case risk is $1/3$. Thus, the mixture g, which is entirely outside $\mathcal{F}$ has a better worst-case risk. We leave it as an exercise to show that in this particular example, the new g is indeed minimax optimal, that is, its risk $R(g,\mathcal{F})$ is equal to $\inf_f R(f,\mathcal{F})$. However, the broader question of the derivation of lower bounds for $R(\mathcal{F})$ requires new methods. The methods and examples are both fun and insightful, and it is not surprising that so many researchers have fallen for the charms of minimax lower bounds. However, rare are the occasions in which we can actually exhibit a minimax optimal estimate.

Before we develop methods for computing lower bounds for the minimax risk, two more points need to be made. The first point concerns attempts at picking estimates by maximum likelihood methods. Consider for example $\mathcal{F}$, the class of all uniform densities on $[0,a]$, with $a > 0$. We observe a single X. If g is in $\mathcal{F}$, then it must be uniform on $[0, r(X)]$ for some function $r(X)$. We note that if the true density has parameter a, then the expected error is

$$\frac{2}{a}\int_0^a \left(1 - \min\left(\frac{r(y)}{a}, \frac{a}{r(y)}\right)\right) dy.$$

As $r(y)$ should clearly grow in proportion to y (why?), we set $r(y) = cy$ for some $c > 1$. Work out the expression and obtain a worst-case risk that is independent of a and thus needs no further maximizing over densities in $\mathcal{F}$:

$$R(g,\mathcal{F}) = 2 - \frac{1}{c} - \frac{2\log c}{c}.$$

This is minimal for $c = \sqrt{e}$, with minimal value $2 - 2/\sqrt{e}$. Thus, using this strategy, it seems that we must estimate by $g(x, X)$, the uniform density on $[0, X\sqrt{e}]$. In the exercises, you will be asked to exhibit a nonuniform g that has a better worst-case risk. Consider now the obvious g one would think of first, if restricted to $\mathcal{F}$, the maximum likelihood choice: it is the density in $\mathcal{F}$ that maximizes $f(X)$ among all $f \in \mathcal{F}$. Clearly, this occurs for the uniform density on $[0, X]$ (the choice $c = 1$). This yields $R(g,\mathcal{F}) = 1$, a clearly suboptimal choice. In fact, in many, but not all cases, the maximum likelihood choice within $\mathcal{F}$ is far from the minimax-optimal solution.

Our second point concerns huge classes. If $\mathcal{F}$ consists of all uniform densities on $[a, b]$, with $a < b$ unrestricted, it is easy to see that $R(\mathcal{F}) = 2$ with one data point. Just note that we might as well assume that $X = 0$. Whatever the form then of $g(x, 0)$, for every $\epsilon > 0$, we find c so large that $\int_{-c}^{c} g(x,0)\,dx > 1 - \epsilon$. Then pick a uniform density on $[-c/\epsilon, c/\epsilon]$. For this density, we have $\int_{-c}^{c} f < \epsilon$. Thus, the total variation distance between g and f is at least $1 - 2\epsilon$, and the L_1 distance is thus at least $2 - 4\epsilon$. As ϵ is arbitrary, we are done. Such classes are hopeless, at least for the sample size given to us (one). It is imperative that we distinguish between these hopeless classes, which are too rich to handle with the given sample size, and smaller manageable classes.

§15.2. The General Minimax Problem

Assume now that we have an i.i.d. sample of size n, $X_1, \ldots, X_n$, drawn from an unknown density $f \in \mathcal{F}$. Then the worst-case risk for a density estimate g_n (a mapping from $\mathbf{R}^d \times (\mathbf{R}^d)^n$ to $\mathbf{R}$) is

$$R(g_n,\mathcal{F}) = \sup_{f\in\mathcal{F}} \mathbf{E}\left\{ \int |g_n(x; X_1,\ldots,X_n) - f(x)|\,dx \right\}.$$

The minimax risk for $\mathcal{F}$ is

$$R_n(\mathcal{F}) = \inf_{g_n} R(g_n,\mathcal{F}).$$

A density estimate is minimax optimal for n if $R(g_n,\mathcal{F}) = R_n(\mathcal{F})$. As such estimates are very hard to describe explicitly, we define

$$C(\{g_n\}) = \limsup_{n\to\infty} \frac{R(g_n,\mathcal{F})}{R_n(\mathcal{F})}$$

and say that $C(\{g_n\})$ is the asymptotic minimax constant for the sequence of estimates $\{g_n\}$. If this is finite, then $\{g_n\}$ is minimax rate optimal.

Assume that we have a density estimate f_n with super-adaptive capabilities in the form of the existence of a constant $\alpha > 0$ and a sequence $\beta_n > 0$ such that for all f, not just those in a given class,

$$\begin{aligned} &\mathbf{E}\left\{\int |f_n(x; X_1, \ldots, X_n) - f(x)|\,dx\right\} \\ &\quad \le \alpha \inf_{g_n \in \mathcal{G}_n} \mathbf{E}\left\{\int |g_n(x; X_1, \ldots, X_n) - f(x)|\,dx\right\} + \beta_n, \end{aligned}$$

where $\mathcal{G}_n$ is typically a very simple class of density estimates that may or may not include the original, possibly complicated, f_n. For example, f_n could be a kernel estimate with data-dependent choice for the bandwidth, while $\mathcal{G}_n$ would then be the class of kernel estimates with the same kernel but fixed bandwidth depending upon n and f only. This is precisely the setting of our results. Note in particular that when f is explicitly known, then the data are not needed, and the choice of a bandwidth as a function of n and f is therefore much easier to deal with. In particular, the worst-case risk for f_n is now bounded as follows:

$$\begin{aligned} R(f_n, \mathcal{F}) &= \sup_{f \in \mathcal{F}} \mathbf{E}\left\{\int |f_n(x; X_1, \ldots, X_n) - f(x)|\,dx\right\} \\ &\le \beta_n + \alpha \sup_{f \in \mathcal{F}} \inf_{g_n \in \mathcal{G}_n} R(g_n, \{f\}) \\ &\le \beta_n + \alpha \inf_{g_n \in \mathcal{G}_n} R(g_n, \mathcal{F}). \end{aligned}$$

To continue the discussion, if we take $\mathcal{F}$ as all decreasing densities on $[0,1]$ bounded by B, then it is well known that the choice $h = 1/(n^{1/3}B^{2/3})$ for the bandwidth in the kernel estimate has a worst-case risk bounded by $C(B/n)^{1/3}$ for a universal constant C. If we use the data-dependent choice for h presented in Chapter 11, and call that estimate f_n, then we have

$$R(f_n, \mathcal{F}) \le C'\sqrt{\log n/n} + 4C(B/n)^{1/3} \le C''(B/n)^{1/3}$$

for other universal constants C' and C''. Neither C nor C' nor C'' depend upon B. While on the surface, this seems to be worse, note however that this inequality is valid for all B, and that f_n does not require the knowledge of B. The estimate f_n thus adapts very well then to all situations. The other option, to first estimate B and then to use that estimated value in the formula $h = 1/(n^{1/3}B^{2/3})$ is of course valid, but totally unnecessary. In other words, if we have simple good bounds for the performance of the kernel estimate when the density is known, then f_n basically comes to within a constant factor of that bound.

§15.3. Rich Classes

We call a class rich if $\liminf_{n\to\infty} R_n(\mathcal{F}) > 0$. The reason why such classes are difficult to handle is because there are too many densities and too few data points. To prove that a class is rich, we merely exhibit for each n a finite subclass of $\mathcal{F}$ and show that if each member of the subclass is equally likely, then regardless of the density estimate, the expected L_1 error over this class does not tend to zero with n.

Partition $\mathbf{R}^d$ first into sets A_i, and assign probabilities p_i to the A_i's, with $\sum_i p_i = 1$. On A_i, find two nonnegative functions g_i, h_i such that

$$\int g_i = \int h_i = p_i,$$

and denote $q_i = \int |g_i - h_i|$. Let θ be a bit vector of finite or infinite dimension with ith component θ_i. Define a subclass of $\mathcal{F}$ parametrized by θ with typical member

$$f_\theta(x) = \begin{cases} g_i(x) & \text{if } x \in A_i \text{ and } \theta_i = 1\ , \\ h_i(x) & \text{if } x \in A_i \text{ and } \theta_i = 0\ . \end{cases}$$

In other words, the bit θ_i is used to choose between g_i and h_i on A_i. Picture these f_θ's as living on the vertices of a hypercube; the fact that this hypercube is a subset of $\mathcal{F}$ means of course that $\mathcal{F}$ is huge. More importantly, the problem has become "combinatorial."

THEOREM 15.1. *If $\mathcal{F}$ is a class of densities containing a subclass of the type described above, then*

$$R_n(\mathcal{F}) \ge \sum_i \frac{q_i(1-p_i)^n}{2}.$$

PROOF. The proof uses the following construction: for each i, consider n i.i.d. random variables $Y_{1i}, \ldots, Y_{ni}$ drawn from the density g_i/p_i, and n i.i.d. random variables $Z_{1i}, \ldots, Z_{ni}$ drawn from the density h_i/p_i, and let all samples be independent. Consider furthermore two other independent random variables or vectors, $\sigma = (\sigma_1, \ldots, \sigma_n)$, a uniform random permutation of $(1, ..., n)$, and $(N_1, N_2, \ldots)$, a multinomial random vector with parameters $(n; p_1, p_2, \ldots)$. We simultaneously describe (very dependent) samples of n i.i.d. random variables for all f_θ's. First of all, each sample has N_i data points in A_i. If $\theta_i = 1$, then draw the N_i observations Y_{ji} with $\sum_{k=1}^{i-1} N_k < j \le \sum_{k=1}^{i} N_k$ from g_i/p_i (a density on A_i). If $\theta_i = 0$, then draw the observations Z_{ji} with $\sum_{k=1}^{i-1} N_k < j \le \sum_{k=1}^{i} N_k$ from h_i/p_i. Finally, order the returned observations, and apply the random permutation σ to them. Observe that this way for each θ we obtain an i.i.d. sample drawn from f_θ. Let Θ be a uniform random bit vector with ith component Θ_i. Let $f_{n,\theta}$ be any density estimate, where θ denotes the fact that we use the sample for θ described above. We argue as follows:

$$
\begin{aligned}
\sup_{f\in\mathcal{F}} \mathbf{E}&\left\{\int |f_n - f|\right\} \\
&\geq \sup_{\theta} \mathbf{E}\left\{\int |f_{n,\theta} - f_\theta|\right\} \\
&\geq \mathbf{E}\left\{\int |f_{n,\Theta} - f_\Theta|\right\} \\
&\geq \sum_i \int_{A_i} \mathbf{E}\{|f_{n,\Theta,i+} - g_i|\mathbf{1}_{[\Theta_i=1]} + |f_{n,\Theta,i-} - h_i|\mathbf{1}_{[\Theta_i=0]}\} \\
&\qquad \text{(where } f_{n,\Theta,i+}, f_{n,\Theta,i-} \text{ use the samples for } \Theta \text{ with the } i\text{th} \\
&\qquad \text{component forced to 1 and 0, respectively)} \\
&\geq 1/2 \sum_i \int_{A_i} \mathbf{E}\{\mathbf{1}_{[N_i=0]}(|f_{n,\Theta,i+} - g_i| + |f_{n,\Theta,i+} - h_i|)\} \\
&= 1/2 \sum_i \int_{A_i} \mathbf{E}\{\mathbf{E}\{\mathbf{1}_{[N_i=0]}(|f_{n,\Theta,i+} - g_i| + |f_{n,\Theta,i-} - h_i|)|\Theta\}\} \\
&\geq 1/2 \sum_i \int_{A_i} \mathbf{E}\{\mathbf{E}\{\mathbf{1}_{[N_i=0]}|g_i - h_i||\Theta\}\} \\
&\qquad \text{(by the fact that on } N_i = 0, \text{ given } \Theta,\ f_{n,\Theta,i+} = f_{n,\Theta,i-}) \\
&= 1/2 \sum_i \int_{A_i} \mathbf{E}\{\mathbf{1}_{[N_i=0]}|g_i - h_i|\} \\
&= 1/2 \sum_i \mathbf{P}\{N_i = 0\} \int_{A_i} |g_i - h_i| \\
&= 1/2 \sum_i \int_{A_i} (1 - p_i)^n q_i.
\end{aligned}
$$

This completes the proof. □

At first sight, the lower bound seems to tend to zero exponentially with n, but one should not forget the freedom in the choice of the p_i's. For example, if we can pick all g_i's and h_i's disjoint, and all A_i's of equal probability mass $p_i = 1/k$, $1 \leq i \leq k$, then $q_i = 2p_i$, and the lower bound becomes $(1 - 1/k)^n \geq 1 - n/k$. Clearly, if k is unrestricted, then we have

$$R_n(\mathcal{F}) \geq 1.$$

Consider the class of all densities on $[0,1]$ bounded by 2. Clearly, we may partition $[0,1]$ into k equal intervals A_i, and if we denote the left and right halves of A_i by A_i' and A_i'', then we may set $g_i = 2\mathbf{1}_{A_i'}$, $h_i = 2\mathbf{1}_{A_i''}$. By the remarks made above, $R_n(\mathcal{F}) \geq 1$. In fact, by considering the dumb estimate $g_n \equiv \mathbf{1}_{[0,1]}$, we see that $\sup_{f\in\mathcal{F}} \int |g_n - f| \leq 1$, and thus, we have $R_n(\mathcal{F}) = 1$. The minimax optimal estimate is thus one that ignores the data!

Another rich class is the class of all densities with infinitely many absolutely continuous derivatives, and $|f^{(s)}| < C_s$ for all integers $s \geq 0$. Not all arbitrary sequences of constants C_s are possible. We just assume that the constants are such that the class has at least one compact support member g_0. In the construction of a subclass for Theorem 1, we let all g_i's and h_i's be translates of g_0, setting

$$g_i(x - c_i) = p_i g_0(x), \; h_i(x - d_i) = p_i g_0(x),$$

for some suitable constants c_i, d_i so that all g_i's and h_i's are disjoint. Note that $f_\theta \in \mathcal{F}$ because

$$g_i^{(s)}(x - c_i) = p_i g_0^{(s)}(x).$$

In a third example, consider all monotone densities on $[0, \infty)$ that are bounded by 1. The lower bound of Theorem 1 is valid for this case too. Consider the set $A_1 = [0, 1]$, and an arbitrary integer $k > 1$. On A_1, define functions g and h as follows: $g = 1$ on $[0, 1/(2k)]$, $g = 1/(2k - 1)$ on $[1/(2k), 1]$ (so that $\int_{A_1} g = 1/k$). Set $h = 1/k$ on $[0, 1]$, so that its integral too is $1/k$. Note that $\int_{A_1} |g - h| = (1 - 1/k)(1/k)$. Now define $A_2 = [1, 1 + (2k - 1)]$, and define a pair (g, h) as on A_1, only shifted by 1 and scaled down by $2k - 1$. Continuing in this manner until A_k, we obtain our subfamily of $\mathcal{F}$ of size 2^k. In Theorem 1, all p_i's are $1/k$ and all q_i's are $1/k - 1/k^2$. Thus,

$$R_n(\mathcal{F}) \geq \frac{(1 - 1/k)(1 - 1/k)^n}{2} \geq \frac{1 - (n + 1)/k}{2}$$

which by the arbitrary nature of k is as close to $1/2$ as desired. Therefore, $R_n(\mathcal{F}) \geq 1/2$ for all n.

We conclude from this partial list of rich classes that no uniform rate of convergence results are possible under tail conditions alone (see first example), under smoothness conditions alone (see second example), and under monotonicity conditions alone (see third example). It is not difficult to see that the third example can be modified so that smoothness and monotonicity conditions alone again do not suffice to get minimax errors converging to zero. The same is true under any combination of tail conditions and monotonicity conditions (by playing on the infinite peak at the origin).

§15.4. Assouad's Lemma

In this section, we develop information-theoretic methods to bound $R(\mathcal{F})$ from below, and base our treatment on the work of Assouad (1983), Bretagnolle and Huber (1979) and Birgé (1986). Assouad's lemma (Theorem 2 below) is based on a discretization argument. Assume that there is a partition

$A_0, A_1, \ldots, A_r$ of $\mathbf{R}^d$ and that there are functions g_{ij}, $0 \le i \le r, j \in \{0,1\}$, such that the hypercube of 2^r functions f_θ, $\theta = (\theta_1, \ldots, \theta_r) \in \{0,1\}^r$, defined by

$$f_\theta(x) = \begin{cases} g_{00}(x) & (x \in A_0)\ , \\ g_{i\theta_i}(x) & (x \in A_i)\ , \end{cases}$$

is included in $\mathcal{F}$. A subset $\mathcal{H}$ of such 2^r functions is called a hypercube of $\mathcal{F}$, for short. The idea is to carefully select $\mathcal{H}$ to optimize a lower bound for the minimax risk. Clearly,

$$R(\mathcal{F}) \ge R(\mathcal{H})$$

so that we need only study bounds for a finite collection of densities organized in a hypercube.

We define two parameters for a given hypercube. The Hellinger closeness is defined by

$$H = \inf_\theta \inf_{1 \le i \le r} \int \sqrt{f_\theta f_{\zeta_i}},$$

where ζ_i agrees with θ in all bits, but has its ith bit opposite to that of θ. By the Cauchy–Schwarz inequality, it is easy to see that $H \le 1$. Furthermore, $H = 1$ when $f_\theta = f_{\zeta_i}$ for all i, that is, all 2^r densities collapse to one density. Thus, when the hypercube consists of densities that are close to each other, H is large. Hence the name, Hellinger closeness.

The second parameter is the L_1 separation of a hypercube,

$$L = \inf_{1 \le i \le r} \int_{A_i} |g_{i0} - g_{i1}| = \inf_\theta \inf_{1 \le i \le r} \int_{A_i} |f_\theta - f_{\zeta_i}|.$$

Clearly, L is large when all densities in the hypercube are sufficiently far from each other on each of $A_1, \ldots, A_r$. The following lemma is due to Assouad (1983) and Birgé (1983).

THEOREM 15.2 (ASSOUAD'S LEMMA). *Let $\mathcal{H}$ be a hypercube of densities f_θ, with $|\mathcal{H}| = 2^r$, with Hellinger closeness $H \ge \beta > 0$, and with L_1 separation $L \ge \alpha > 0$. Then*

$$R_n(\mathcal{H}) \ge \begin{cases} \frac{r\alpha}{2}\left(1 - \sqrt{2 - 2\beta^n}\right), \\ \frac{r\alpha}{2}\left(1 - \sqrt{2n(1-\beta)}\right), \\ \frac{r\alpha\beta^{2n}}{4}. \end{cases}$$

PROOF. The middle bound follows from the first one, because $2 - 2(1 - (1-\beta))^n \le 2 - 2(1 - n(1-\beta)) = 2n(1-\beta)$. We will derive a lower bound that is valid for any density estimate f_n. We write $\mathcal{X}_n$ for $(x_1, \ldots, x_n)$. Let θ_{ik} agree with θ in all bits but force its ith bit to be k, $k \in \{0,1\}$. Observe the following:

$$
\begin{aligned}
&R_n(f_n, \mathcal{H})\\
&\quad = \sup_\theta \mathbf{E}\left\{\int |f_n - f_\theta|\right\}\\
&\quad \ge \mathbf{E}\left\{\int |f_n - f_\Theta|\right\}\\
&\qquad \text{(where } \Theta \text{ is uniform on } \{0,1\}^n)\\
&\quad = \frac{1}{2^r}\sum_\theta \int \left(\int |f_n(x, \mathcal{X}_n) - f_\theta(x)|\,dx\right) \prod_{j=1}^n f_\theta(x_j) \prod_{j=1}^n dx_j\\
&\quad = \frac{1}{2^{r+1}}\sum_\theta \left(\sum_{i=1}^r \int\int_{A_i} |f_n(x, \mathcal{X}_n) - f_{\theta_{i0}}(x)|\,dx \prod_{j=1}^n f_{\theta_{i0}}(x_j) \prod_{j=1}^n dx_j\right.\\
&\qquad\qquad \left. + \sum_{i=1}^r \int\int_{A_i} |f_n(x, \mathcal{X}_n) - f_{\theta_{i1}}(x)|\,dx \prod_{j=1}^n f_{\theta_{i1}}(x_j) \prod_{j=1}^n dx_j\right)\\
&\quad \ge \frac{1}{2^{r+1}}\sum_\theta \int \sum_{i=1}^r \left(\int_{A_i} |f_n(x, \mathcal{X}_n) - f_{\theta_{i0}}(x)|\,dx\right.\\
&\qquad\qquad \left. + \int_{A_i} |f_n(x, \mathcal{X}_n) - f_{\theta_{i1}}(x)|\,dx\right)\\
&\qquad \times \min\left(\prod_{j=1}^n f_{\theta_{i0}}(x_j), \prod_{j=1}^n f_{\theta_{i1}}(x_j)\right) \prod_{j=1}^n dx_j\\
&\quad \ge \frac{1}{2^{r+1}}\sum_\theta \int \left(\int |f_{\theta_{i1}}(x) - f_{\theta_{i0}}(x)|\,dx\right)\\
&\qquad \times \min\left(\prod_{j=1}^n f_{\theta_{i0}}(x_j), \prod_{j=1}^n f_{\theta_{i1}}(x_j)\right) \prod_{j=1}^n dx_j\\
&\quad \ge \frac{r\alpha}{2^{r+1}}\sum_\theta \int \min\left(\prod_{j=1}^n f_{\theta_{i0}}(x_j), \prod_{j=1}^n f_{\theta_{i1}}(x_j)\right) \prod_{j=1}^n dx_j\\
&\quad \ge \frac{r\alpha}{2}\inf_{\theta,i} \int \min\left(\prod_{j=1}^n f_{\theta_{i0}}(x_j), \prod_{j=1}^n f_{\theta_{i1}}(x_j)\right) \prod_{j=1}^n dx_j.
\end{aligned}
$$

When f and g are two densities, LeCam's inequality (Theorem 5.4) states that $2\int \min(f,g) \geq \left(\int \sqrt{fg}\right)^2$. Using this with f and g replaced by product densities, we see that

$$\begin{aligned} R_n(f_n, \mathcal{H}) &\geq \frac{r\alpha}{4} \inf_{\theta,i} \left(\int \sqrt{\prod_{j=1}^n f_{\theta_{i0}}(x_j) f_{\theta_{i1}}(x_j)} \prod_{j=1}^n dx_j \right)^2 \\ &\geq \frac{r\alpha}{4} \left(\inf_{\theta,i} \int \sqrt{f_{\theta_{i0}}(x_j) f_{\theta_{i1}}(x_j)}\, dx_j \right)^{2n} \\ &\geq \frac{r\alpha\beta^{2n}}{4}. \end{aligned}$$

This proves the third bound. For two densities f, g, we also have, since the L_1 distance is not more than twice the Hellinger distance (see Exercise 5.7 and Section 5.8):

$$\int \min(f,g) = 1 - \frac{\int |f-g|}{2} \geq 1 - \sqrt{\int (\sqrt{f} - \sqrt{g})^2} = 1 - \sqrt{2 - 2\int \sqrt{fg}}$$

which, when applied to the product densities we are working with yields the first bound. □

Applying Assouad's lemma boils down to the identification of a hypercube in which the L_1 distance between neighbors is large, yet the Hellinger distance between neighbors is not too large. Several examples follow in the next two sections and in the exercises.

§15.5. Example: The Class of Convex Densities

In this section we work out the details of the following particular example. Let $\mathcal{F}_B$ be the class of nondecreasing, convex-shaped densities f on $[0,1]$ with $\sup_{[0,1]} f(x) \leq B$. For convenience we consider the concrete value $B = 7/3$. Then we have the following:

LEMMA 15.1. *For $n \geq 288$, we have*

$$R_n(\mathcal{F}_{7/3}) \geq \frac{1}{37.37192\ldots n^{2/5}}.$$

PROOF. Let r be a suitable positive integer to be determined further on. We will construct a hypercube family of 2^r densities contained in $\mathcal{F}_{7/3}$ satisfying the conditions of Assouad's lemma. The interval $[0,1]$ is partitioned into r intervals denoted by $A_1, \ldots, A_r$. Thus, $A_i = [(i-1)/r, i/r)$. On $[(i-1)/r, i/r]$, we consider two piecewise linear functions f_i and g_i with the following properties:

A. $f_i = g_i$ at the ends of the interval, and $f_{i-1}((i-1)/r) = f_i((i-1)/r)$;
B. $\int_{A_i} f_i = \int_{A_i} g_i$;

C.

$$\sup_{x\in((i-2)/r,(i-1)/r)} \max(f'_{i-1}(x), g'_{i-1}(x)) \leq \inf_{x\in((i-1)/r,i/r)} \min(f'_i(x), g'_i(x));$$

D. $\sup_{x\in A_{i-1}} \max(f_{i-1}(x), g_{i-1}(x)) \leq \inf_{x\in A_i} \min(f_i(x), g_i(x))$;
E. $f_1(0) = g_1(0) = A > 0$, where $A = 1/3$;
F. $f_r(1) = g_r(1) \leq 7/3$;
G. $\sum_{i=1}^r \int_{A_i} f_i = 1$.

If we piece together a function f by choosing either f_i or g_i on A_i, then we obtain a bona fide density (by Parts B, D, E and G). Furthermore, f is increasing (by Part D), continuous on (0,1) (by Part A) and convex (by Parts A and C). There are 2^r such possible functions, and we may parametrize the family by a bit vector $\theta = (\theta_1, \ldots, \theta_r)$, where $\theta_i = 1$ if the corresponding f picks f_i on A_i, and $\theta_i = 0$ otherwise. We will denote this f by f_θ.

Define $d_i = 4ai$ for some constant a. We will pick the f_i's such that on A_i, both f'_i and g'_i are between d_{i-1} and d_i. This will then insure nondecreasing derivatives, and thus convexity for f. The positivity is also insured. On A_i, we set

$$f'_i = \begin{cases} d_{i-1} + a, & x \in [(i-1)/r, (i-1/3)/r), \\ d_i, & x \in [(i-1)/r + 2/(3r), i/r), \end{cases}$$

and

$$g'_i = \begin{cases} d_{i-1} & x \in [(i-1)/r, (i-2/3)/r), \\ d_i - a = d_{i-1} + 3a, & x \in [(i-1)/r + 1/(3r), i/r). \end{cases}$$

It takes a moment to verify that $\int_{A_i} f'_i = \int_{A_i} g'_i$ so that f_i and g_i make equal jumps. Another routine computation shows that $\int_{A_i} f_i = \int_{A_i} g_i$ as well. Thus, the functions will suit us, provided that the total integral is one. To do this, we observe that

$$\int_{A_i} f'_i = \frac{2}{3r}(4a(i-1) + a) + \frac{4ai}{3r} = \frac{4ai - 2a}{r},$$

and similarly for $\int_{A_i} g'_i$. Hence, at i/r, the value of any f in our class is

$$A + \sum_{j=1}^{i} \frac{4aj - 2a}{r} = A + \frac{2ai^2}{r}.$$

For future reference, we note that $f_r(1) = g_r(1) = A + 2ar$. Next, we compute the integrals:

$$\int_{A_i} f_i = \int_0^{2/3r} (f_i((i-1)/r) + (4(i-1)a + a)x)\, dx$$

$$+\int_0^{1/3r} (f_i(i/r) - 4aix)\,dx$$
$$= \frac{A}{r} + \frac{a(18i^2 - 18i + 22)}{9r^2}.$$

Thus, the integral of any f in our class is the sum over the r individual pieces, which is

$$A + \frac{a(6r^2 + 16)}{9r}.$$

As $A = 1/3$ and the integral must be one, we must take

$$a = \frac{6r}{6r^2 + 16}.$$

We verify quickly that $a \le 1/r$, and that for $r \ge 2$, $a \ge 3/(5r)$. Finally, we insure that each f is bounded by $7/3$:

$$f_r(1) = 1/3 + \frac{12r^2}{6r^2 + 16} = \frac{42r^2 + 16}{18r^2 + 48} < 7/3.$$

This ends the construction of our parametric family with 2^r members.

To apply Assouad's lemma, we need to compute two lower bounds α and β, where α is a uniform lower bound on $\int_{A_i} |f_i - g_i|$, and β is a uniform lower bound on $\int \sqrt{fg}$, where f and g are two of the 2^r functions that differ only on one interval. The required uniformity is with respect to i, the interval index. Clearly,

$$\int_{A_i} |f_i - g_i| = 2\left(\int_0^{1/(3r)} ax\,dx + \int_0^{1/(6r)} 2ax\,dx\right) = \frac{a}{6r^2}.$$

Thus, we may set

$$\alpha = \frac{a}{6r^2}.$$

Next, pick i, and let $I = A_i$, $O = [0,1] - A_i$. Let f and g be two of our functions that are equal on all intervals except A_i. Note that on A_i, $|f_i - g_i|$ is maximal at $1/3r$ from the left boundary, and the value there is $a/(3r)$. Then since

$$\int \sqrt{fg} = 1 - 1/2 \int (\sqrt{f} - \sqrt{g})^2$$

and

$$\begin{aligned}\int (\sqrt{f} - \sqrt{g})^2 &= \int \frac{(f-g)^2}{(\sqrt{f} + \sqrt{g})^2} \\ &\le 3/4\,(f-g)^2 \\ &\qquad (\text{since } f, g \ge 1/3) \\ &= 3/4\,(f-g)^2\end{aligned}$$

$$= {}^3\!/\!_2\left(\int_0^{1/(3r)} (ax)^2\,dx + \int_0^{1/(6r)} (2ax)^2\,dx\right)$$

$$= \frac{a^2}{36r^3},$$

we have

$$\int \sqrt{fg} \geq 1 - \frac{a^2}{72r^3} \geq 1 - \frac{1}{72r^5}.$$

We set $1-\beta = 1/(72r^5)$. By Assouad's lemma,

$$\begin{aligned}\inf_{f_n}\sup_{f\in\mathcal{F}_{7/3}} \mathbf{E}\left\{\int |f_n - f|\right\} &\geq \frac{r\alpha}{2}\left(1-\sqrt{2n(1-\beta)}\right)\\ &= \frac{ra}{12r^2}\left(1-\sqrt{\frac{n}{36r^5}}\right)\\ &\geq \frac{1}{20r^2}\left(1-\sqrt{1/4}\right)\\ &\qquad\text{(provided } r\geq 2 \text{ and } r^5 \geq n/9)\\ &= \frac{1}{40r^2}\\ &= \frac{1}{40\lceil (n/9)^{1/5}\rceil^2}\\ &\qquad\text{(upon taking } r = \lceil (n/9)^{1/5}\rceil)\\ &\geq \frac{1}{40(1+(n/9)^{1/5})^2}.\end{aligned}$$

Note that the condition $r \geq 2$ holds if $n \geq 288$. In that case, $1 \leq (n/9)^{1/5}/2$, and the lower bound is thus at least

$$\frac{1}{40((3/2)(n/9)^{1/5})^2} \geq \frac{1}{37.37192\ldots n^{2/5}}$$

which was to be shown. □

§15.6. Additional Examples

One worked-out example of the use of Assouad's lemma suffices for illustration. In this section, we discuss some other examples, but defer all details to the exercise section.

In a first example is given a central density f^*, and define $\mathcal{F}_\epsilon = \{f : \int |f - f^*| < \epsilon\}$, where $\epsilon \in (0,1)$. In the exercises, you will be guided to show that $R_n(\mathcal{F}_\epsilon) \geq \epsilon$. Because $\sup_{f\in\mathcal{F}_\epsilon} \int |f - f^*| \leq \epsilon$, the estimator $f_n = f^*$ is minimax-optimal for the class $\mathcal{F}_\epsilon$. Thus, if minimizing the minimax criterion is our goal, then there is no need to consider the data; the data can in fact be thrown away, and we can take $f_n = f^*$.

Consider next a Lipschitz class of the simplest kind. We say that a density f is Lipschitz with constant C on a set S when it has a version such that $|f(x) - f(y)| \leq C|x - y|$ for all $x, y \in S$. Let $\mathcal{L}$ be the class of all Lipschitz C densities on $[0,1]$. In the exercises, hints are given to show that $R_n(\mathcal{L}) \geq (aC/n)^{1/3}$ for some universal constant a and all n and C. Also, this bound is best possible for all n and C up to a multiplicative constant by taking a kernel density estimate with uniform kernel and bandwidth $h = b/(C^2 n)^{1/3}$ where b is another universal constant b (see the exercises of Chapter 9). In this definition of h, the constant C must be known. However, we may also apply the data-dependent kernel bandwidth method of Chapter 11 and get the expected L_1 error rate that is bounded by a constant times $(C/n)^{1/3}$, without of course knowing C!

Consider the class of all densities of random variables X distributed as NY where N is a normal $(0,1)$ random variable and Y is an arbitrary random variable. When $1/Y$ is normal, then NY is Cauchy. When $1/Y$ is distributed as $2E$, E standard exponential, then NY is Laplace. And when $1/Y^2$ is gamma $(a/2)$ divided by $2a$, then NY is Student with parameter a. This is clearly a huge class. Observe that $\log|X|$ is distributed as $\log|N| + \log|Y|$. Thus, after transformation, the class may be considered as a given convolution class for a fixed random variable Z with a density, that is, the class of all densities of random variables of the form $Z+Y$, where Y is arbitrary and independent of Z. Convolution classes are harder to handle than the classes seen so far since member densities cannot conveniently be constructed on partitions of the space. Nevertheless, it is possible to prove that all convolution classes are rich, their minimax risks being at least 1 (see the exercises).

§15.7. Tuning the Parameters of Variable Kernel Estimates

The purpose of this section is to show that it is impossible to find a close-to-optimal way of selecting the smoothing parameter of the kernel estimate if the smoothing parameter is allowed to depend on the location. More precisely, consider the class of variable kernel estimates

$$f_{n,h(x)}(x) = \frac{1}{n}\sum_{i=1}^{n} K_{h(x)}(x - X_i),$$

where K is a fixed kernel on $\mathbf{R}$, and the smoothing parameter $h(x)$ is allowed to be any measurable function $h : \mathbf{R} \to (0,\infty)$ of x. Various ways of selecting $h(x)$ in a data-dependent manner have been proposed, so that the obtained data-based variable kernel estimate has the form

$$f_{n,H(x)}(x) = \frac{1}{n}\sum_{i=1}^{n} K_{H(x)}(x - X_i),$$

where it is understood that $H(x) = H(x; X_1, \ldots, X_n)$. Ideally, one would like to find $H(x)$ so that the expected error $\mathbf{E}\{\int |f_{n,h(x)}(x) - f(x)|\,dx\}$ is close to the ideal value $\inf_{h:\mathbf{R}\to(0,\infty)} \mathbf{E}\{\int |f_{n,h(x)}(x) - f(x)|\,dx\}$ for all densities. The next theorem shows that it is impossible even if f is restricted to be a smooth bounded function of a compact support.

As in Section 5, let $\mathcal{F}_B$ be the class of nondecreasing, convex-shaped densities f on $[0,1]$ with $\sup_{(0,1)} f(x) \le B$. We state the result for the simplest possible kernels. Introduce a shape parameter for nonnegative symmetric kernels K:

$$\rho \stackrel{\text{def}}{=} \text{support}(K) \times \int K^2.$$

Note that by the Cauchy–Schwarz inequality, $\rho \ge 1$, and equality is reached for the uniform kernel. In any case, ρ is a scale-invariant parameter that will appear in our bounds.

THEOREM 15.3. *Let K be a symmetric nonnegative square-integrable kernel with shape parameter ρ. Then, for $n \ge 288$:*

$$\inf_{H:\mathbf{R}^{n+1}\to(0,\infty)} \sup_{f\in\mathcal{F}_{7/3}} \frac{\mathbf{E}\{\int |f_{n,H(x)}(x) - f(x)|\,dx\}}{\inf_{h:\mathbf{R}\to(0,\infty)} \mathbf{E}\{\int |f_{n,h(x)}(x) - f(x)|\,dx\}} \ge \frac{n^{1/10}}{41\sqrt{\rho}}.$$

The theorem shows that even with the knowledge that $f \in \mathcal{F}_B$, one cannot efficiently design a variable bandwidth. It is generally accepted that some versions of variable bandwidths outperform fixed bandwidth estimates, and the theorem above does not contradict this. What it says is that the class of variable bandwidth kernel estimates is too large to be optimized. One should constrain the class of possible functions $h(x)$ from which selection is made. In Section 12.3 we show that if we partition the real line into k intervals and use a different (fixed) bandwidth on each interval, then simultaneous optimization of the k intervals and the k bandwidths is possible in the sense that there exists a data-based choice of a piecewise constant function $H(x)$ with k pieces such that for all densities

$$\mathbf{E}\left\{\int |f_{n,H(x)}(x) - f(x)|\,dx\right\}$$
$$\le c_0 \inf \int |f_{n,h(x)}(x) - f(x)|\,dx + c_1\sqrt{\frac{k\log n}{n}},$$

where c_0 and c_1 are absolute constants, and the infimum is taken over all variable kernel estimates where $h(x)$ is piecewise constant with k pieces.

To prove Theorem 3, we combine two results, an estimate of the L_1 error for raw variable kernel estimates inspired by the work of Sain and Scott (1997), and the minimax lower bound for $\mathcal{F}_B$ shown in Section 5. Sain and Scott remarked that for most locations x, the kernel estimate could be made unbiased by taking $h(x)$ fixed and positive (not depending upon n). A similar but less explicit remark may also be found in Hazelton (1996, p. 223). This prompted us to consider $f \in \mathcal{F}_B$, where every x has this property. In particular, the rate of convergence could be $O(1/\sqrt{n})$ if we knew f and thus the optimal map $h : \mathbf{R} \to (0, \infty)$.

LEMMA 15.2. *Let K be a symmetric nonnegative square-integrable kernel on the real line with shape parameter ρ. Then*

$$\sup_{f \in \mathcal{F}_B} \inf_{h:\mathbf{R}\to(0,\infty)} \mathbf{E}\left\{\int |f_{n,h(x)}(x) - f(x)|\, dx\right\} \le \sqrt{\frac{4B\rho}{n}}.$$

PROOF. Fix $f \in \mathcal{F}_B$. Without loss of generality, let the support of K be $[-1, 1]$, so that $\rho = 2\int K^2$. Note that we may set $h(x) = |x|$ for $x < 0$ and $h(x) = x - 1$ for $x > 1$, so that since f has support on $[0, 1]$ and K on $[-1, 1]$, we have $f_{n,h(x)} \equiv f(x) \equiv 0$ outside $[0, 1]$. Thus, pick $x \in (0, 1)$. Note that as long as $h = h(x)$ is less than $\min(x, 1 - x)$:

$$\begin{aligned}
\mathbf{E}\{f_{n,h}(x)\} &= f * K_h(x) \\
&= (1/h) \int f(y) K((x - y)/h)\, dy \\
&= \int f(x + hw) K(w)\, dw \\
&= \mathbf{E}\{f(x + hW)\} \quad \text{(where } W \text{ has density } K) \\
&\ge f(x + h\mathbf{E}\{W\}) \quad \text{(by Jensen's inequality)} \\
&= f(x).
\end{aligned}$$

Also, it is trivial to check that $\lim_{h\to\infty} f * K_h(x) = 0$. Thus, since $f * K_h(x)$, considered as a function of h, is continuous in h, we see that necessarily $f(x) = f * K_h(x)$ for some $h > 0$, and, in fact, there exists at least one such h with $h \ge \min(x, 1 - x)$. For $x \in (0, 1)$, we define

$$h(x) = \sup\{z > 0 : f * K_z(x) = f(x)\},$$

and note for further use that $h(x) \ge \min(x, 1 - x)$. From here on, h denotes this choice. Then, by routine calculations,

$$\begin{aligned}
\mathbf{E}\{|f_{n,h}(x) - f(x)|\} &= \mathbf{E}\{|f_{n,h}(x) - \mathbf{E}f_{n,h}(x)|\} \\
&\le \sqrt{\mathbf{E}\{(f_{n,h}(x) - \mathbf{E}f_{n,h}(x))^2\}} \\
&= \sqrt{\frac{\mathbf{V}\{K_h(x - X_1)\}}{n}} \\
&\le \sqrt{\frac{\mathbf{E}\{(K_h(x - X_1))^2\}}{n}}
\end{aligned}$$

$$= \sqrt{\frac{\int K^2((x-y)/h)h^{-2}f(y)\,dy}{n}}$$
$$\leq \sqrt{\frac{B\int K^2}{nh}}.$$

This is true for all $x \in (0,1)$, and integration yields

$$\mathbf{E}\left\{\int |f_{n,h}(x) - f(x)|\,dx\right\} \leq \sqrt{\frac{B\int K^2}{n}} \int_0^1 \frac{1}{\sqrt{h(x)}}\,dx$$
$$\leq 2\sqrt{\frac{Bc\int K^2}{n}} \int_0^{1/2} \frac{1}{\sqrt{x}}\,dx$$
$$= \sqrt{\frac{4B\rho}{n}}$$

as required. □

It is of course impossible to estimate all f's in $\mathcal{F}_B$ at the rate $1/\sqrt{n}$ with any density estimate, let alone the data-based variable kernel density estimate. To support this, we merely require the minimax lower bound from Section 5. We may now easily prove the theorem.

PROOF OF THEOREM 3. By Lemmas 1 and 2:

$$\inf_{H:\mathbf{R}^{n+1}\to(0,\infty)} \sup_{f\in\mathcal{F}_{7/3}} \frac{\mathbf{E}\{\int |f_{n,H(x)}(x) - f(x)|\,dx\}}{\inf_{h:\mathbf{R}\to(0,\infty)} \mathbf{E}\{\int |f_{n,h(x)}(x) - f(x)|\,dx\}}$$
$$\geq \frac{\inf_{f_n} \sup_{f\in\mathcal{F}_{7/3}} \mathbf{E}\{\int |f_{n,H(x)}(x) - f(x)|\,dx\}}{\sup_{f\in\mathcal{F}_{7/3}} \inf_{h:\mathbf{R}\to(0,\infty)} \mathbf{E}\{\int |f_{n,h(x)}(x) - f(x)|\,dx\}}$$
$$\geq \frac{1/37.37192\ldots n^{2/5}}{\sqrt{28\rho/3n}}$$
$$\geq \frac{n^{1/10}}{41\sqrt{\rho}}.$$

This concludes the proof of Theorem 3. □

§15.8. Sufficient Statistics

There is yet a different method for obtaining lower bounds, which uses the fact that the best estimators depend upon sufficient statistics for the family only, and that the properties of simple sufficient statistics are often well known. As an example, we take the class

$$\mathcal{F} = \{f \mid f = pg + (1-p)h,\ p \in [0,1]\},$$

where g and h are known densities with disjoint supports. Other examples of simple parametric families may be found in the exercises.

THEOREM 15.4. *Let f_n be any density estimate with $\int f_n = 1$. Then*

$$R(f_n, \mathcal{F}) \geq \frac{\gamma_n}{\sqrt{n}},$$

where $\inf_{n \geq 4} \gamma_n \geq 0.030153$, *and* $\gamma_n \to 0.0849\ldots$ *as* $n \to \infty$.

PROOF. Define $p_n = \int_A f_n$ where A is the support of g. We first claim that

$$g_n = p_n g + (1 - p_n) h$$

is at least as good an estimate as f_n. This follows from

$$\begin{aligned} \int |g_n - f| &= \int_A g|p_n - p| \ + \ \int_{A^c} h|p_n - p| \\ &= 2|p_n - p| \\ &= 2\left|\int_A (f_n - f)\right| \\ &\leq \int |f_n - f|. \end{aligned}$$

Next, we note that N, the cardinality of A, is a sufficient statistic for p. Thus, we should be able to reduce the data to N. This can be done by invoking the conditional form of Jensen's inequality:

$$\mathbf{E}\left\{\int |f_n - f|\right\} \geq 2\mathbf{E}\{|p_n - p|\} \geq 2\mathbf{E}\{|\mathbf{E}\{p_n|N\} - p|\} = 2\mathbf{E}\{|\psi(N) - p|\},$$

where ψ is some measurable function, which can be considered as an estimator based upon N only. The dependence upon n is suppressed temporarily. Observe that the joint density of the data given N is

$$\frac{1}{n!} \sum_\sigma \prod_{i=1}^N f(x_{\sigma_i}) \prod_{j=N+1}^n g(x_{\sigma_j}),$$

where $\sigma = (\sigma_1, \ldots, \sigma_n)$ is a permutation of $(1, \ldots, n)$. This density is independent of p. This is another way of stating that in this problem, N is a sufficient statistic for p. This is why $\mathbf{E}\{p_n|N\}$ is $\psi(N)$ and not $\psi_p(N)$. Now, we can randomize p, by making p equal to $1/2$ with probability $1/2$, and equal to $1/2 + c/\sqrt{n}$ with probability $1/2$. Let N_1, N_2 be binomial random variables with parameters $(n, 1/2)$ and $(n, 1/2 + c/\sqrt{n})$ respectively. Then

$$\begin{aligned} &\sup_p \mathbf{E}_p\left\{\int |f_n - f|\right\} \\ &\quad\geq \tfrac{1}{2}\mathbf{E}_{1/2}\left\{\int |f_n - f|\right\} \ + \ \tfrac{1}{2}\mathbf{E}_{1/2+c/\sqrt{n}}\left\{\int |f_n - f|\right\} \\ &\quad\geq \mathbf{E}\{|\psi(N_1) - \tfrac{1}{2}|\} + \mathbf{E}\{|\psi(N_2) - \tfrac{1}{2} - c/\sqrt{n}|\} \end{aligned}$$

$$\geq \sum_{j \leq n/2} \left(|\psi(j) - \tfrac{1}{2}| \mathbf{P}\{N_1 = j\} + |\psi(j) - \tfrac{1}{2} - c/\sqrt{n}| \mathbf{P}\{N_2 = j\} \right)$$
$$\geq \sum_{j \leq n/2} \left(|\psi(j) - \tfrac{1}{2}| + |\psi(j) - \tfrac{1}{2} - c/\sqrt{n}| \right) \mathbf{P}\{N_2 = j\}$$
$$\geq \frac{c}{\sqrt{n}} \mathbf{P}\{N_2 \leq n/2\}.$$

Now, let Z be binomial with parameters n and $\tfrac{1}{2} + c/\sqrt{n}$. Then

$$\mathbf{P}\{Z \leq n/2\} = \mathbf{P}\left\{ \frac{Z - \mathbf{E}Z}{\sqrt{\mathbf{V}\{Z\}}} \leq \frac{-c\sqrt{n}}{\sqrt{n(\tfrac{1}{2} + c/\sqrt{n})(\tfrac{1}{2} - c/\sqrt{n})}} \right\}$$
$$\to \Phi(-2c),$$

where Φ is the normal distribution function. Here we tacitly used the central limit theorem and a continuity argument. Thus, our lower bound is asymptotic to

$$\frac{c\Phi(-2c)}{\sqrt{n}}$$

and has the largest possible value when $2c = 0.7517915241\ldots$. □

§15.9. Bibliographic Remarks

Surveys of recent results may be found in Devroye (1987), Devroye and Györfi (1985), Yang and Barron (1999), with relevant L_1 material also available in the work of Birgé (1985) and Bretagnolle and Huber (1979). All results cited in the text for monotone densities on the real line are in Devroye (1987). For the monotone densities on $[0, 1]$ bounded by B, the asymptotic minimax rate as a joint function of B and n decreases as $(\log B/n)^{1/3}$. This was shown by Birgé via the histogram estimate (1987a) and Grenander's estimate (Birgé, 1989). Additional results were obtained by Groeneboom (1985, 1986). The minimax rates for very smooth densities have been studied by Konakov (1972), Ibragimov and Khasminskii (1982), and Devroye (1987). The example of the L_1 ball centered around f^* is from Devroye (1987), based on a similar result for Hellinger balls by Birgé (1985). The basic hypercube methodology for lower bounds was developed in a series of papers by Assouad (1983) and Birgé (1983, 1985, 1986). A rival methodology based on simplices via Fano's inequality was put in the spotlight by Yu (1997), Yang (1996), and Yang and Barron (1999). Rate of convergence results that are asked to pick the worst density out of $\mathcal{F}$ before the sample size n is revealed are dealt with in a series of papers by Devroye (1983, 1995) and Birgé (1986).

Various aspects of the variable kernel estimate have been studied by Breiman, Meisel, and Purcell (1977), Habbema, Hermans, and Remme (1978), Abramson (1982), Devroye (1985), Devroye and Penrod

(1986), Hall and Marron (1988), Mielniczuk, Sarda, and Vieu (1989), Terrell and Scott (1992), Hall (1992), Marron, Hall, and Hu (1995), Sain and Scott (1996), and Hazelton (1996). Lemma 2 is inspired by interesting observations about unbiased estimates summarized in Sain and Scott (1997) (see also Hazelton, 1996). Theorem 3 also appears in Devroye and Lugosi (2000).

§15.10. Exercises

Exercise 15.1 (MONOTONE DENSITIES). Let f be a monotonically nonincreasing density on $[0, \infty)$, and let g be a density estimate. Then explicitly construct a monotonically nonincreasing density estimate g^* with $\int |g^* - f| \leq \int |g - f|$, $g^* \geq 0$, $\int g^* = 1$. (Thus, decreasing densities are best estimated by decreasing densities.)

Exercise 15.2 (MULTIVARIATE MONOTONE DENSITIES). Repeat the previous exercise for densities on $(\mathbf{R}^+)^d$ that are decreasing in every coordinate.

Exercise 15.3 (UNIFORM DENSITIES). Let $\mathcal{F}$ be the class of uniform densities on $[0, a], a > 0$. Assume that only one observation X is available.

A. Show that the best density estimate g restricted to $\mathcal{F}$ must be the uniform density on cX for some constant $c > 1$. Conclude that the uniform density on $[0, X\sqrt{e}]$ is best possible within this restricted class.
B. Find a density estimate $g(x, X)$ with worst-case risk $R(g, \mathcal{F}) < 2 - 2/\sqrt{e}$, thereby beating the best estimate restricted to $\mathcal{F}$. HINT: By the previous exercises, one may clearly restrict oneself to nonincreasing densities.

Exercise 15.4 (CONVEX CLASSES). A convex class of densities is such that if f and g are in the class, then so are $\lambda f + (1 - \lambda)g$ for all $\lambda \in (0, 1)$. A closed class is such that all L_1 limits of sequences of densities from the class are in the class, provided that the limits are densities. Let $\mathcal{F}$ be a closed convex class of densities. Let f_n be a density estimate, and assume that it is a density. Let g_n be its projection on $\mathcal{F}$, defined by

$$g_n = \arg\min_{f \in \mathcal{F}} \int |g_n - f_n|.$$

A. Prove that g_n exists (i.e., the minimum is attained).
B. Prove that for any $f \in \mathcal{F}$, $\int |g_n - f| \leq \int |f_n - f|$, so the projection estimate is an improvement, regardless of which density in f is being estimated. In other words, densities in such classes are best estimated by their own kind.

C. Show that $\sup_{f\in\mathcal{F}} \int |g_n - f| \le \sup_{f\in\mathcal{F}} \int |f_n - f|$.
D. Show that the following classes are closed and convex: the nonincreasing densities on $[0,\infty)$; the convex densities on $[0,\infty)$; the class of all densities on $\mathbf{R}^d$; the scale mixture class for a fixed density f on the real line, that is, if X has density f, then this class contains all densities of ZX, where Z is any random variable independent of X; the convolution class for a fixed density f on the real line, that is, if X has density f, then this class contains all densities of $Z+X$, where Z is any random variable independent of X.

Exercise 15.5 (THE DECISIVE ESTIMATE). When $\mathcal{F} = \{f_1, f_2\}$, and only one observation X is available, we defined the decisive estimate $g(x,y)$ as follows: $g(x,y) = f_1(x)\mathbf{1}_A(y) + f_2(x)\mathbf{1}_{A^c}(y)$ for a set A. Show that

$$R(g,\mathcal{F}) = \max\left\{\int_A f_2, \int_{A^c} f_1\right\} \times \int |f_1 - f_2|$$

and that this is minimal for a set B nested between the Scheffé sets $A = \{f_1 > f_2\}$ and $A^* = \{f_1 \ge f_2\}$.

Exercise 15.6 (THE HESITATOR). Continuing the previous exercise, the hesitator forms a mixture of f_1 and f_2 based on X, so that $g(x,y) = f_1(x)\lambda(y) + f_2(x)(1-\lambda(y))$ for some function $\lambda : \mathbf{R}^d \to [0,1]$. This exercise explores the worst-case risk of g.

A. Show that $R(g,\mathcal{F}) = \int |f_1 - f_2| \times \max(\int f_1(1-\lambda), \int f_2\lambda)$.
B. With the choice $\lambda = f_1/(f_1+f_2)$, verify that $R(g,\mathcal{F}) = \int |f_1 - f_2| \times \int f_1 f_2/(f_1+f_2)$. Show that the last integral is $\int \min(f_1,f_2) - \int \min^2(f_1,f_2)/(f_1+f_2)$.
C. By a clever use of the Cauchy–Schwarz inequality, show that $R(g,\mathcal{F}) \le J(1-J^2/4)/2$, where $J = \int |f_1 - f_2|$.
D. Compare the hesitator with the decisive estimate.

Exercise 15.7 Let f_1 and f_2 be uniform on $[0,1]$ and $[1/2, 1]$, respectively. Show that the minimax optimal estimator for one data point X is as follows: if $X \in [0, 1/2]$, set $g = f_1$; if $X \in (1/2, 1]$, set $g = 5/3\mathbf{1}_{[1/2,1]}(x) + 1/3\mathbf{1}_{[0,1/2]}(x)$.

Exercise 15.8 Prove that $R_n(\mathcal{F}_\epsilon) = \epsilon$, where $\mathcal{F}_\epsilon = \{f : \int |f - f^*| < \epsilon\}$, $\epsilon \in (0,1)$. Do this by using Assouad's lemma, and following these steps. Let r be large but fixed, and partition the real line into r intervals of equal f^*-probability $1/r$. Form the hypercube of 2^r densities as follows: split the ith set into two equiprobable intervals, B_i and C_i, and consider a piece of density that is $(1+\epsilon)f^*$ on B_i and $(1-\epsilon)f^*$ on C_i, and another density that has $1+\epsilon$ and $1-\epsilon$ switched. Compute the Hellinger closeness and the L_1 separation, use these in the first of Assouad's inequalities, and let r tend to infinity.

Exercise 15.9 (CONTINUATION). Prove that $R_n(\mathcal{F}_\epsilon) = \epsilon$, where $\mathcal{F}_\epsilon = \{f : \int |f - f^*| = \epsilon\}$, $\epsilon \in (0,1)$. Show that the minimax-optimal estimate is $f_n = f^*$. Show also that the best estimate f_n restricted to the class has $R_n(f_n, \mathcal{F}_\epsilon) = 2\epsilon$.

Exercise 15.10 (HELLINGER BALLS). Consider the class of densities $\mathcal{F}_\epsilon = \{f : H_2(f, f^*) \le \epsilon\}$, where f^* is a given fixed density, $H_2(f, f^*) = \sqrt{\int(\sqrt{f} - \sqrt{f^*})^2}$ is the Hellinger distance, and $\epsilon \in (0,1)$ is fixed. Determine the minimax risk and the minimax optimal estimate. HINT: You may want to use the previous two exercises, and some inequalities linking H_2 with the L_1 distance.

Exercise 15.11 (L_∞ BALLS). Consider the class of densities

$$\{f : L_\infty(f, f^*) \le \epsilon\},$$

where f^* is a fixed density and $\epsilon > 0$ is given, and L_∞ denotes the supremum difference between the Lebesgue versions $L(f)$ and $L(g)$ of f and g, respectively. Show that he minimax risk is $\int \min(f, \epsilon)$.

Exercise 15.12 Prove that the minimax risk for the class of all bounded densities is 2.

Exercise 15.13 Let $\mathcal{L}$ be the class of all Lipschitz C densities on $[0,1]$. Via Assouad's lemma, show that $R_n(\mathcal{L}) \ge (aC/n)^{1/3}$ for some universal constant a and all n and C by following these steps. Partition $[0,1]$ into r equal intervals. On each interval, consider two functions that begin and end with value 1, and have value 1 in the middle. These functions are translates of the following splines: spline one connects $(0,1)$ with $(1/4r, 1 + C/4r)$, $(3/4r, 1 - C/4r)$ and $(1/r, 1)$. Spline two connects $(0,1)$ with $(1/4r, 1 - C/4r)$, $(3/4r, 1 + C/4r)$ and $(1/r, 1)$. Define the hypercube of 2^r densities based upon these splines. Show that the L_1 separation is of the order of C/r^2. Compute a good lower bound for the Hellinger closeness, and optimize Assouad's lower bound with respect to r.

Exercise 15.14 Let $\mathcal{L}$ be the class of all densities f on $[0,1]$ satisfying the following property: $|f(x) - f(y)| \le C|x - y|^\alpha$, $x, y \in [0,1]$, where $\alpha \in (0,1]$ and $C > 0$ are fixed. Via Assouad's lemma, show that

$$R_n(\mathcal{L}) \ge \left(\frac{aC}{n^\alpha}\right)^{1/(1+2\alpha)}$$

for some universal constant a depending upon α only, and all n and C. HINT: Generalize the previous exercise.

Exercise 15.15 Let $\mathcal{L}$ be the class of all densities f on $[0,1]$ satisfying the following property: $|f^{(s)}(x) - f^{(s)}(y)| \le C|x-y|^\alpha$, $x, y \in (0,1)$, where s is a positive integer, and $\alpha \in (0,1]$ and $C > 0$ are fixed. Furthermore, on $(0,1)$, f has $s-1$ absolutely continuous derivatives. Via Assouad's lemma, show that

$$R_n(\mathcal{L}) \ge \left(\frac{aC}{n^{s+\alpha}}\right)^{1/(1+2(s+\alpha))}$$

for some universal constant a depending upon s and α only, and all n and C. HINT: Generalize the previous exercise.

Exercise 15.16 Show that all convolution classes are rich by applying Assouad's lemma. Recall that convolution classes have all distributions of $Z+Y$, Z fixed, Y arbitrary and independent of Z. Follow these steps. Let f be the density of the defining random variable Z in the convolution class. First find M so large that $[-M, M]$ captures most of the mass of f. Build a hypercube by choosing $|Y|$ equiprobable on $\{1, 2, \ldots, r\}$, and by setting $Y = |Y|\theta_{|Y|}$, where θ takes all 2^r values on $\{-1,1\}^r$. The densities in the hypercube should be virtually nonoverlapping. Set up an appropriate partition for Assouad's lemma, and show that the minimax risk is $1 - o(1)$ as $r \to \infty$.

Exercise 15.17 Consider the family of all densities of the form $\sum_{i=1}^k p_i g_i$, where the p_i's form a probability vector, and the g_i's are (possibly overlapping) densities. The g_i's are known, but the p_i's are not. Derive a minimax lower bound for this class in terms of n, k and possibly the distances between the g_i's.

Exercise 15.18 (THE NORMAL DENSITIES). Consider the class $\mathcal{F}$ of all zero mean normal densities.

A. Prove that $R_n(\mathcal{F}) \ge c/\sqrt{n}$ for some positive constant c.
B. Construct a minimax-optimal estimate f_n. (Hard.)
C. For $n = 1$, what would your estimate look like, if it were (i) the maximum likelihood estimate; (ii) the estimate within $\mathcal{F}$ of the smallest worst-case risk; (iii) the minimax-optimal estimate?

Exercise 15.19 (BOUNDED SUPPORT CHARACTERISTIC FUNCTIONS). Let $\mathcal{F}$ be the class of all densities whose characteristic function has compact support. Show that this class is rich. HINT: Use the inequality $\int |f-g| \ge \sup_t |\varphi(t) - \psi(t)|$, where ϕ and ψ are the characteristic functions corresponding to f and g.

Exercise 15.20 (BOUNDED SUPPORT CHARACTERISTIC FUNCTIONS). Let $\mathcal{F}$ be the class of all densities whose characteristic function has support on $[-T, T]$. Is this class rich?

Exercise 15.21 (CONCAVE DENSITIES). Prove that for the class of all concave densities on $[0, 1]$, $R_n(\mathcal{F}) \geq cn^{-2/5}$ for some universal constant $c > 0$. Can this bound be improved?

Exercise 15.22 (CONVEX DENSITIES).

A. Show that the class of all densities that are convex on $[0, \infty)$ and bounded by one is rich.
B. Show that the subclass supported on $[0, 2T]$ has a minimax risk that tends to zero, where $T > 1$ is a fixed constant. Determine the rate of decrease to zero of $R_n(\mathcal{F})$.
C. Determine $R_n(\mathcal{F})$ for the further subclass consisting of all convex densities that are continuous on $[0, \infty)$, bounded by one, nonincreasing, supported on $[0, 2T]$ for some $T > 1$, and whose derivative is bounded in absolute value by another constant D, which is suitably large. (The boundedness of the derivative will cause a different answer.)
D. Show that the class of all densities that are convex and supported on $[0, 1]$ is rich.

Exercise 15.23 (NORMALIZED LOG-CONCAVE DENSITIES). Let $\mathcal{F}$ be the class of all densities for which $\log f$ is concave, $f(0) = 1$, and a mode occurs at 0 (as log-concave densities are necessarily unimodal). Show that $\mathcal{F}$ is totally bounded, and that $R_n(\mathcal{F}) = \Theta(n^{-2/5})$.

Exercise 15.24 (THE BETA DENSITIES). Let $\mathcal{F}$ be the class of beta densities on $[0, 1]$. Show that $R_n(\mathcal{F}) = \Theta(1/\sqrt{n})$.

Exercise 15.25 (AKHIEZER CLASSES). An Akhiezer class of order r (a nonnegative integer) consists of densities on $[0, 1]$ that have r absolutely continuous derivatives on $[0, 1]$, denoted by $f^{(0)}, f^{(1)}, \ldots, f^{(r)}$. It is characterized by polarities (bits) $b_0, b_1, \ldots, b_r$, and positive real numbers $B_0, B_1, \ldots, B_r$ where $f^{(i)}$ is nonincreasing on $[0, 1]$ if $b_i = 1$ and nondecreasing there if $b_i = 0$. Furthermore, the maximal absolute value of $f^{(i)}$ on $[0, 1]$ is not larger than B_i. The class with $r = 0$, $b_0 = 1$ and any value $B_0 \geq 1$ is the class of all monotonically nonincreasing functions on $[0, 1]$. For the convex nonincreasing functions, we add $b_1 = 0$, and a suitably big value for B_1. Determine the rate of decrease to zero for the minimax rate for any r (assuming that the constants are all suitably big).

Exercise 15.26 Let $\mathcal{F}$ be a class of uniformly bounded densities. Let f_n

be a density estimate, where n is a fixed integer. Show that if

$$\lim_{m\to\infty} \sqrt{m} R_m(\mathcal{F}) = \infty$$

then either f_n is not unbiased (i.e., $\sup_{f\in\mathcal{F}} \int |\mathbf{E} f_n - f| > 0$) or

$$\sup_{f\in\mathcal{F}} \int \sqrt{\mathbf{E}\{f_n^2\}} = \infty.$$

In other words, information for finite n follows from an asymptotic statement about the class.

§15.11. References

I. Abramson, "On bandwidth variation in kernel estimates—a square root law," *Annals of Statistics*, vol. 10, pp. 1217–1223, 1982.

P. Assouad, "Deux remarques sur l'estimation," *Comptes Rendus des Séances de l'Académie des Sciences. Série I*, vol. 296, pp. 1021–1024, 1983.

L. Birgé, "Approximation dans les espaces métriques et théorie de l'estimation," *Zeitschrift für Wahrscheinlichkeitstheorie und verwandte Gebiete*, vol. 65, pp. 181–237, 1983.

L. Birgé, "Non-asymptotic minimax risk for Hellinger balls," *Probability and Mathematical Statistics*, vol. 5, pp. 21–29, 1985.

L. Birgé, "On estimating a density using Hellinger distance and some other strange facts," *Probability Theory and Related Fields*, vol. 71, pp. 271–291, 1986.

L. Birgé, "On the risk of histograms for estimating decreasing densities," *Annals of Statistics*, vol. 15, pp. 1013–1022, 1987a.

L. Birgé, "Estimating a density under order restrictions: Nonasymptotic minimax risk," *Annals of Statistics*, vol. 15, pp. 995–1012, 1987b.

L. Birgé, "The Grenander estimator: A nonasymptotic approach," *Annals of Statistics*, vol. 17, pp. 1532–1549, 1989.

L. Birgé and P. Massart, "Rates of convergence for minimum contrast estimators," *Probability Theory and Related Fields*, vol. 97, pp. 113–150, 1993.

L. Breiman, W. Meisel, and E. Purcell, "Variable kernel estimates of multivariate densities," *Technometrics*, vol. 19, pp. 135–144, 1977.

J. Bretagnolle and C. Huber, "Estimation des densités: Risque minimax," *Zeitschrift für Wahrscheinlichkeitstheorie und verwandte Gebiete*, vol. 47, pp. 119–137, 1979.

L. Devroye, "On arbitrarily slow rates of global convergence in density estimation," *Zeitschrift für Wahrscheinlichkeitstheorie und verwandte Gebiete*, vol. 62, pp. 475–483, 1983.

L. Devroye, "A note on the L_1 consistency of variable kernel estimates," *Annals of Statistics*, vol. 13, pp. 1041–1049, 1985.

L. Devroye, *A Course in Density Estimation*, Birkhäuser-Verlag, Boston, 1987.

L. Devroye, "Another proof of a slow convergence result of Birgé," *Statistics and Probability Letters*, vol. 23, pp. 63–67, 1995.

L. Devroye and L. Györfi, *Nonparametric Density Estimation: The L_1 View*, Wiley, New York, 1985.

L. Devroye and G. Lugosi, "Variable kernel estimates: On the impossibility of tuning the parameters," in: *High-Dimensional Probability* (edited by E. Giné and D. Mason), Springer-Verlag, New York, 2000.

L. Devroye and C. S. Penrod, "Distribution-free lower bounds in density estimation," *Annals of Statistics*, vol. 12, pp. 1250–1262, 1984.

P. Groeneboom, "Estimating a monotone density," in: *Proceedings of the Berkeley Conference in Honor of Jerzy Neyman and Jack Kiefer, Vol. II* (edited by L. Le Cam and R. A. Olshen), pp. 539–555, Wadsworth, Belmont, CA., 1985.

P. Groeneboom, "Some current developments in density estimation," in: *CWI Monograph 1, Mathematics and Computer Science* (edited by J. W. de Bakker, M. Hazewinkel and J. K. Lenstra), pp. 163–192, Elsevier Science, Amsterdam, 1986.

J. D. F. Habbema, J. Hermans, and J. Remme, "Variable kernel density estimation in discriminant analysis," in: *COMPSTAT 78* (edited by L. C. A. Corsten and J. Hermans,), Physica Verlag, Wien, 1978.

P. Hall, "On global properties of variable bandwidth density estimators," *Annals of Statistics*, vol. 20, pp. 762–778, 1992.

P. Hall and J. S. Marron, "Variable window width kernel estimates," *Probability Theory and Related Fields*, vol. 80, pp. 37–49, 1988.

M. Hazelton, "Bandwidth selection for local density estimation," *Scandinavian Journal of Statistics*, vol. 23, pp. 221–232, 1996.

I. A. Ibragimov and R. Z. Khasminskii, "Estimation of distribution density belonging to a class of entire functions," *Theory of Probability and its Applications*, vol. 27, pp. 551–562, 1982.

V. D. Konakov, "Non-parametric estimation of density functions," *Theory of Probability and its Applications*, vol. 17, pp. 361–362, 1972.

J. S. Marron, P. Hall, and T. C. Hu, "Improved variable window estimators of probability densities," *Annals of Statistics*, vol. 23, pp. 1–10, 1995.

J. Mielniczuk, P. Sarda, and P. Vieu, "Local data-driven bandwidth choice for density estimation," *Journal of Statistical Planning and Inference*, vol. 23, pp. 53–69, 1989.

S. R. Sain and D. W. Scott, "On locally adaptive density estimation," *Journal of the American Statistical Association*, vol. 91, pp. 1525–1534, 1996.

S. R. Sain and D. W. Scott, "Zero-bias locally adaptive density estimators," Technical Report, Rice University, Houston, 1997.

G. R. Terrell and D. W. Scott, "Variable kernel density estimation," *Annals of Statistics*, vol. 20, pp. 1236–1265, 1992.

Y. Yang, "Minimax optimal density estimation," PhD Dissertation, Yale University, 1996.

Y. Yang and A. Barron, "Information-theoretic determination of minimax rates of convergence," *Annals of Statistics*, vol. 27, pp. 1564–1599, 1999.

B. Yu, "Assouad, Fano and Le Cam," in: *Festschrift for Lucien Le Cam* (edited by D. Pollard, E. Torgersen, and G. Yang), pp. 423–435, Springer-Verlag, New York, 1997.

Chapter 16
Choosing the Kernel Order

§16.1. Introduction

Using the notation $K_h(\cdot)$ to denote $(1/h)K(\cdot/h)$, let the univariate kernel estimate with bandwidth $h > 0$ and kernel K be

$$f_{n,K,h}(x) = \frac{1}{n}\sum_{i=1}^{n} K_h(x - X_i).$$

The problem dealt with in this chapter is that of the joint choice of the pair (h, K). When restricted to positive kernels, we know that

$$\liminf_{n\to\infty} n^{2/5} \inf_{h>0,K\geq 0} \mathbf{E}\left\{\int |f_{n,K,h} - f|\right\} \geq 0.86$$

and we know that this lower bound rate can be achieved for a big class of smooth small-tailed densities and for all symmetric bounded kernels that have compact support. In fact, the Epanechnikov kernel is asymptotically best possible, and there seems to be little point in the joint choice of h and $K \geq 0$. However, when we allow negative-valued kernels, the picture changes, and in fact, for every density f and for every n, there is an optimal pair (h, K). Unfortunately, picking both without restrictions of any kind is much too hard. If K is fixed but possibly not nonnegative, then we define

$$\Phi(n, K) = \inf_f \inf_{h>0} \mathbf{E}\left\{\int |f_{n,K,h} - f|\right\}.$$

The best rate of convergence depends upon K. Densities f for which $\inf_{h>0} \mathbf{E}\{\int |f_{n,K,h} - f|\} = O(\Phi(n, K))$ are said to be in the saturation class for K.

For an even positive integer s, introduce the class of kernels $\mathcal{K}_s$ consisting of all even kernels L for which $\int L = 1$, $\int x^i L(x)\,dx = 0$ for $1 \leq i < s$, $\nu_s \stackrel{\text{def}}{=} \int |x|^s |L(x)|\,dx < \infty$, and $\mu_s \stackrel{\text{def}}{=} |\int x^s L(x)\,dx| \neq 0$, and for which the smallest unimodal majorant (the smallest unimodal function dominating $|L|$) is in L_1 and L_4. Note that most standard positive kernels are in $\mathcal{K}_2$. We give several examples of kernels from $\mathcal{K}_s$ in the exercises. For all kernels in $\mathcal{K}_s$, the best possible rates of convergence are identical, $n^{-s/(2s+1)}$. We mention the following result without proof.

THEOREM 16.1 (DEVROYE, 1988). *Let f be a density with characteristic function φ. Let $f_{n,L,h}$ be the standard kernel estimate and let its kernel L belong to $\mathcal{K}_s$ for some even s. If L has Fourier transform ψ, then*

$$\liminf_{n\to\infty} n^{s/(2s+1)} \inf_{h>0} \mathbf{E}\left\{\int |f_{n,L,h} - f|\right\} \geq A(f)B(L),$$

where

$$A(f) \stackrel{\text{def}}{=} \left(\int \sqrt{f}\right)^{2s/(2s+1)} \left(\sup_t |t|^s |\varphi(t)|\right)^{1/(2s+1)}$$

and

$$B(L) \stackrel{\text{def}}{=} \left(\sqrt{\int \psi^2 / 2\sqrt{8\pi}}\right)^{2s/(2s+1)} (\mu_s/s!)^{1/(2s+1)}.$$

This inequality remains formally valid even if $A(f) = \infty$. Also, for any density f,

$$A(f) \geq (\pi(s-1)/s)^{s/(2s+1)},$$

and for all $L \in \mathcal{K}_s$, $B(L) \geq \left(1/2\sqrt{8\pi}\right)^{2s/(2s+1)} (\mu_s/2\nu_s)^{1/(2s+1)}$.

$A(f)$ is an appropriate measure of the "difficulty" posed by f for kernels $L \in \mathcal{K}_s$. It is composed of a tail factor, $\int \sqrt{f}$, and a smoothness factor depending upon the behavior of φ. It can be verified that $A(f)$ is scale and translation invariant. Its value depends upon the shape of the density only. It is infinite for one of two reasons only, either because the tail is too large, or the density is not smooth enough. For nonnegative kernels and $s = 2$, we obtain, in particular,

$$\liminf_{n\to\infty} n^{2/5} \mathbf{E}\left\{\int |f_{n,K,h} - f|\right\} \geq 2^{-13/5} = 0.164938488\ldots.$$

As $\mathcal{K}_s$ imposes an absolute lower bound on the convergence rate for all densities, it is interesting to characterize the densities for which the given rate is attained, either individually, or uniformly over a subclass. Sufficient conditions are given in Devroye and Györfi (1985, p. 208) and Devroye (1987) for

$$\limsup_{n\to\infty} n^{s/(2s+1)} \inf_{h>0} \mathbf{E}\left\{\int |f_{n,K,h} - f|\right\} < \infty.$$

It suffices that $\int |x|^{1+\epsilon} f(x)\,dx < \infty$ for some $\epsilon > 0$, that f has $s-1$ absolutely continuous derivatives, that $f^{(s)}$ is continuous, and that $\int |f^{(s)}| < \infty$. But under these conditions, the characteristic function φ of f satisfies the inequality

$$\sup_t |t|^s |\varphi(t)| \leq \int |f^{(s)}| < \infty.$$

In other words, all these densities are in the saturation class for all kernels in $\mathcal{K}_s$. The saturation classes shrink as $s \uparrow \infty$. Interestingly, there are densities, such as the normal density, that are jointly in all saturation classes (see Chapter 17 for more on these special densities). Others end their life in a particular saturation class, say $\mathcal{K}_6$, and thus do not make it to the next class, say $\mathcal{K}_8$. The best choice of h depends on n and s, the highest saturation class index for a given density. So, a joint selection method should be designed to jointly pick the pair (h, K).

Within one class $\mathcal{K}_s$, there is evidence that indicates that one particular kernel is optimal (see the exercises), and thus that it is futile to include many kernels from the same class in the optimization ensemble. So, for data-based selection of the kernel, we include one kernel from each class in our collection of target kernels. The family of kernels will be indexed by the class index, $K_2, K_4, \ldots$, where $K_s \in \mathcal{K}_s$. The objective is to find the optimal pair (h, s) by the minimum distance method of Chapter 10. There are many possible nice families of kernels—we settled on one suggested by Hall and Marron (1988). We will automatically find the best pair from this infinite selection (up to the inevitable factor of three), and this without complexity regularization or penalized model selection (where one gives penalties as s increases). Furthermore, explicit nonasymptotic inequalities will be provided. As a side benefit, this choice is truly adaptive in the minimax sense, that is, our method of defining a kernel estimate gives minimax rates that are jointly optimal (modulo a constant factor) over various subclasses of densities defined by the finiteness of $\int |f^{(s)}|$ and the compactness of the support (thus, without requiring the a priori knowledge of s or a bound on the support or $\int |f^{(s)}|$).

The remainder of this chapter is rather technical. First we discuss the selection of the best out of a finite collection of kernel estimates, each having its own kernel and bandwidth, and then we turn our infinite kernel problem into a finite one to solve the problem at hand.

§16.2. Standard Kernel Estimate: Riemann Kernels

A Borel set A of $\mathbf{R}^d$ is called a star interval if for any $y \in \mathbf{R}^d$, $\{t \in \mathbf{R} : ty \in A\}$ is an interval. Thus, all convex sets are star intervals. Recall from Chapter 11 that a kernel K is said to be Riemann of order k if there exist star intervals $A_1, \ldots, A_k$ and real numbers a_i such that

$$K(x) = \sum_{i=1}^{k} a_i \mathbf{1}_{A_i}(x),$$

where $\mathbf{1}_A$ denotes the indicator function of a set A. We require furthermore that $\int K = 1$. We will call k the Riemann order, which should not be confused with the order of a kernel, which is the smallest positive integer s such that $\int x^s K(x)\,dx \neq 0$, and is in this sense only defined for univariate kernels. When K is fixed and h is chosen by the method of Chapter 11, based on data splitting, then by Lemma 11.1, the following shatter coefficient for the Yatracos class is applicable:

$$S_{\mathcal{A}_\Theta}(m) \le (m+1)(1+2km(n-m))^2 \le 18k^2n^2m^3.$$

Here Θ refers to the parameter space (h in this case), and $\mathcal{A}_\Theta$ is the Yatracos class. Let us widen the scope a bit and pick a Riemann kernel from a finite class of N Riemann kernels, $\mathcal{K} = \{K_1, \ldots, K_N\}$, and choose the bandwidth h simultaneously as well. This is done by formally putting $\Theta = \{(h,j) : h > 0, j \in \{1, \ldots, N\}\}$. Again, Theorem 11.1 applies, but now with a slightly larger shatter coefficient:

LEMMA 16.1. *Consider the class Θ in which $h > 0$ and $K \in \mathcal{K}$ are the free parameters, and assume that all kernels in $\mathcal{K}$ are Riemann of order not exceeding k. Then*

$$S_{\mathcal{A}_\Theta}(m) \le 18k^2n^2m^3N^2.$$

PROOF. Set $r = n - m$. We first consider $N = 2$, and let the kernels in $\mathcal{K}$ be K and L, and assume without loss of generality that their Riemann orders are exactly k. Define the vector

$$z_u = \left(\sum_{i=1}^{r} K\left(\frac{y_1 - X_i}{u}\right), \ldots, \sum_{i=1}^{r} K\left(\frac{y_m - X_i}{u}\right) \right) \in \mathbf{R}^m.$$

As $u \uparrow \infty$, each component of z_u changes every time $(y_j - X_i)/u$ enters or leaves a set A_ℓ, $1 \le \ell \le k$, for some X_i, $1 \le i \le r$, where the A_ℓ's are the star intervals in the definition of K. Note that for fixed $(y_j - X_i)$, the evolution is along an infinite ray anchored at the origin. By our assumption on the possible form of the sets A_ℓ, the number of different values a component can take in its history (as $u \uparrow \infty$) is clearly bounded by $2kr$. As there are m components, the cardinality of the set of different values of z_u is bounded by

$$|\{z_u : u > 0\}| \le 1 + 2kmr.$$

If we define z'_u similarly as z_u, but replace K in the definition by L, then we have

$$\left|\{z'_u : u > 0\}\right| \le 1 + 2kmr$$

as well. Therefore,

$$\left|\{(z_u, z'_v) : u, v > 0\}\right| \le (1 + 2kmr)^2.$$

The same bound applies for the pairs (z_u, z_v), (z'_u, z'_v), and (z'_u, z_v).

Let $\mathcal{W} = \{(w, w') : (w, w') = (z_u, z'_v) \text{ for some } u, v > 0\}$. For fixed $(w, w') \in \mathcal{W}$, let $U_{(w,w')}$ denote the collection of all (u, v) such that $(z_u, z'_v) = (w, w')$. For $(u, v) \in U_{(w,w')}$, we have

$$y_i \in A_{u,v} \quad \text{if and only if} \quad w_i \ge \left(\frac{u}{v}\right)^d w'_i,$$

where w, w' have components w_i, w'_i respectively, $1 \le i \le m$. Thus,

$$|\{\{y_1, \ldots, y_m\} \cap A_{u,v} : (u,v) \in U_{(w,w')}\}| \\ \le \left|\left\{\left(\mathbf{1}_{w_1 \ge cw'_1}, \ldots, \mathbf{1}_{w_m \ge cw'_m}\right) : c \ge 0\right\}\right| \le m+1.$$

But then

$$|\{\{y_1, \ldots, y_m\} \cap A_{u,v} : (u,v) > 0\}| \\ \le (m+1)|U_{(w,w')}| \le (m+1)(1+2kmr)^2.$$

The same bound applies for the three other types of pairs, (z_u, z_v), (z'_u, z'_v), and (z'_u, z_v). Thus,

$$\mathbb{S}_{\mathcal{A}_\Theta}(m) \le 4(m+1)(1+2kmr)^2 \le 8m(3kmr)^2 = 72m^3k^2r^2.$$

If we have a choice between N kernels, we apply the bound not four times, but N^2 times, for all possible pairings (with repetition), to obtain

$$\mathbb{S}_{\mathcal{A}_\Theta}(m) \le N^2(m+1)(1+2kmr)^2 \le 2N^2m(3kmr)^2 = 18N^2m^3k^2r^2. \quad \square$$

Lemma 1 permits us to obtain fine inequalities even when the kernel is freely picked from a finite class. However, in all cases, the kernels have to be Riemann of finite order. In the next section, we deal with the joint selection of h and K for general (non-Riemann) K, and this includes kernels of infinite order.

§16.3. Standard Kernel Estimates: General Kernels

Recall from Chapter 11 that a kernel K is said to be Riemann approximable if for each n there exists a finite number k such that there exists a Riemann kernel K' of order k with

$$\int |K - K'| \le \frac{1}{n}.$$

The smallest such k is called the *kernel complexity* κ_n. If there is a finite class of kernels $K \in \mathcal{K}$, then we need to find Riemann approximations K' for each K individually. A kernel estimate with Riemann kernel K' is piecewise constant and thus easy to work with in simulations.

Define the kernel estimates

$$f_{n-m,K',h}(x) = \frac{1}{n-m} \sum_{i=1}^{n-m} K'_h(x - X_i)$$

for all $h > 0$ and $K \in \mathcal{K}$. Let the pair (H, K) (where K has Riemann approximation K') be selected from $\Theta = (0,\infty) \times \mathcal{K}$ such that

$$\sup_{A \in \mathcal{A}} \left| \int_A f_{n-m,K',h} - \mu_m(A) \right|$$

is minimal, where $\mathcal{A}$ is defined as the collection of all sets

$$\{x : f_{n-m,L',u}(x) \geq f_{n-m,M',v}(x)\}$$

with $u, v > 0$ and L', M' are Riemann approximations of kernels L, M from $\mathcal{K}$. After the selection, the Riemann kernels are no longer needed. Finally, our estimate is $f_{n-m,K,H}$. One may wonder if the deviation via Riemann kernels is really needed. It seems that the combinatorial arguments that will follow may be made to work for certain classes of kernels such as polynomials, but in any case, the generality achieved here will be lost. We offer the following nonasymptotic bound:

THEOREM 16.2. *Consider the kernel density estimate with joint choice of H and K as described above, where K is taken from a class $\mathcal{K}$ of N kernels with kernel complexities uniformly bounded by κ_n. Then for all n, $m \leq n/2$, d, and f,*

$$\begin{aligned}
\mathbf{E}&\left\{\int |f_{n-m,K,H} - f|\right\} \\
&\leq 3 \inf_{h>0, L\in\mathcal{K}} \mathbf{E}\left\{\int |f_{n,L,h} - f|\right\}\left(1 + \frac{2m}{n-m} + 8\sqrt{\frac{m}{n}}\right) \\
&\quad + 8\sqrt{\frac{\log(36N^2\kappa_n^2 m^3 n^2)}{m}} + \frac{30}{n}.
\end{aligned}$$

For n even and $m = n/2$, we thus have

$$\begin{aligned}
\mathbf{E}\left\{\int |f_{n-m,K,H} - f|\right\} \leq &\inf_{h>0, L\in\mathcal{K}} \mathbf{E}\left\{\int |f_{n,L,h} - f|\right\}(9 + 24/\sqrt{2}) \\
&+ 8\sqrt{\frac{\log(9/2\, N^2\kappa_n^2 n^5)}{m}} + \frac{30}{n}.
\end{aligned}$$

PROOF. Note that

$$\mathbf{E}\left\{\int |f_{n-m,K,H} - f|\right\} \leq \mathbf{E}\left\{\int |f_{n-m,K',H} - f|\right\} + \frac{1}{n}.$$

Furthermore,

$$\inf_{h>0, L\in\mathcal{K}} \mathbf{E}\left\{\int |f_{n,L',h} - f|\right\} \leq \inf_{h>0, L\in\mathcal{K}} \mathbf{E}\left\{\int |f_{n,L,h} - f|\right\} + \frac{1}{n}.$$

Thus, a combination with Theorem 10.3 then yields, with the appropriate definition of Θ:

$$\begin{aligned}
\mathbf{E}&\left\{\int |f_{n-m,K,H} - f|\right\} \\
&\leq \mathbf{E}\left\{\int |f_{n-m,K',H} - f|\right\} + \frac{1}{n}
\end{aligned}$$

$$
\begin{aligned}
&\le 3 \inf_{h>0, L\in\mathcal{K}} \mathbf{E}\left\{\int |f_{n,L',h}-f|\right\}\left(1+\frac{2m}{n-m}+8\sqrt{\frac{m}{n}}\right)\\
&\quad + 8\sqrt{\frac{\log(2\mathbb{S}_{\mathcal{A}_\Theta}(m))}{m}}+\frac{4}{n}\\
&\le 3 \inf_{h>0, L\in\mathcal{K}} \mathbf{E}\left\{\int |f_{n,L,h}-f|\right\}\left(1+\frac{2m}{n-m}+8\sqrt{\frac{m}{n}}\right)\\
&\quad + 8\sqrt{\frac{\log(2\mathbb{S}_{\mathcal{A}_\Theta}(m))}{m}}+\frac{30}{n}\\
&\le 3 \inf_{h>0, L\in\mathcal{K}} \mathbf{E}\left\{\int |f_{n,L,h}-f|\right\}\left(1+\frac{2m}{n-m}+8\sqrt{\frac{m}{n}}\right)\\
&\quad + 8\sqrt{\frac{\log(36N^2\kappa_n^2 m^3 n^2)}{m}}+\frac{30}{n},
\end{aligned}
$$

where we used Lemma 1. □

The above inequality is useful when N is finite. We will treat the case of infinite kernel classes in the next section. The other quantity of interest is κ_n. Recall that kernels with $\kappa_n = O(n^b)$ for some finite b are said to be *polynomially Riemann approximable.* All kernels of practical interest are in this class (see Section 11.3). For fixed N, the last two terms in the upper bound of Theorem 2 are then $O(\sqrt{\log n/m})$, just as in the case of Riemann kernels. Obviously, if K is Riemann of order k, then $\kappa_n \le k$. In Chapter 11 several bounds are given for the complexity of popular kernels. What is of interest in the next section, however, is the bound obtained in Section 11.3 for Lipschitz kernels.

We finish this section by noting the impact of Theorem 2 if all N kernels are of orders up to and including an even number s, that is, each of the kernels K is bounded, symmetric, and has finite nonzero sth moment and at least one kernel has zero ith moments for $0 < i < s$. Then regardless of the density and the choice of h,

$$
\liminf_{n\to\infty} n^{s/(2s+1)} \inf_h \mathbf{E}\left\{\int |f_{n,K,h}-f|\right\} > 0
$$

(Theorem 1). For such higher-order kernels, let $m = o(n)$ be such that $m/(n^{2s/(2s+1)}\log n) \to \infty$. Then if $\kappa_n = O(n^\alpha)$ for some finite α, uniformly over the N kernels,

$$
\mathbf{E}\{\int |f_{n,K,H}-f|\} \le (3+o(1)) \inf_{h, L\in\mathcal{K}} \mathbf{E}\{\int |f_{n,L,h}-f|\} + o(n^{-s/(2s+1)}),
$$

and therefore

$$
\sup_f \limsup_{n\to\infty} \frac{\mathbf{E}\left\{\int |f_{n,K,H}-f|\right\}}{\inf_{h,L\in\mathcal{K}} \mathbf{E}\left\{\int |f_{n,L,h}-f|\right\}} \le 3.
$$

Thus, Theorem 2 shows asymptotic optimality to within a factor of 3 for all (finite collections of) kernels of finite order.

§16.4. An Infinite Family of Kernels

We are finally ready to show how one can pick a kernel from an infinite class together with a suitable bandwidth. We consider the univariate class of kernels considered by Hall and Marron in 1988: the kernels have symmetric stable-like characteristic functions given by $\psi(t) = \exp(-|t|^s)$, with s taking positive even integer values. Note that except for $s = 2$, these do not correspond to bona fide densities. This class is interesting and challenging for many reasons. First of all, while $\int K_s = 1$ for all s, we have $\int |K_s| \to \infty$ as $s \to \infty$, because the kernels converge to the sinc kernel, that is, the kernel with characteristic function $\mathbf{1}_{[-1,1]}(t)$. The kernels become increasingly oscillatory as s increases, and the tails of K_s become increasingly heavy. For utility in L_1, the class of kernels should have a limit in L_1, so we modify the Hall–Marron class by considering $\psi_s(t) = (1 - |t|)\exp(-|t|^s)$. Then define $\phi_s(t) = 2\psi_s(t/2) - \psi_s(t)$, and note that for its corresponding kernel K_s, $\int K_s = 1$ for all s and that ϕ_s tends to the trapezoidal characteristic function $\phi_\infty(t) \stackrel{\text{def}}{=} 2(1 - |t|/2)_+ - (1 - |t|)_+$.

Lemma 16.2. *For ϕ_s, with s even and positive, the kernel complexity satisfies*

$$\kappa_n \le 1 + 18432\xi s^2 n^3,$$

where $\xi = (6 + \sqrt{2\pi})/\pi < 2.71$.

Proof. Let K denote K_s, let ϕ denote ϕ_s, and let ψ be ψ_s. As $K(x) = (1/2\pi)\int \cos(tx)\phi(t)\,dt$, we see that

$$\begin{aligned}\sup_x |K'(x)| &\le \frac{1}{\pi}\int_0^\infty t|\phi(t)|\,dt \le \frac{1}{\pi}\left(6 + 2\int_2^\infty te^{-(t/2)^s}\,dt\right)\\ &\le \frac{1}{\pi}\left(6 + 2\int_2^\infty te^{-(t/2)^2}\,dt\right) \le \xi.\end{aligned}$$

Thus, all kernels are Lipschitz with the same Lipschitz constant ξ. Next, let L be the inverse transform for ψ. Then x^2L is the inverse transform for ψ''. Consider $t > 0$. Then

$$\psi''(t) = e^{-t^s}(t^{s-1}(s^2 + s) - t^{s-2}(s^2 - s) - t^{2s-2}s^2).$$

From this, we see that on $[0,1]$, $|\psi''| \le 2s^2t^{s-2}$ and on $[1,\infty)$, $|\psi''| \le 2s^2t^{2s-2}e^{-t^s}$. Using these estimates, we have

$$\sup_x x^2|L(x)| \le \int |\psi''(t)|\,dt \le \frac{4s^2}{s-1} + 4s^2\int_1^\infty t^{2s-2}e^{-t^s}\,dt$$

$$\leq \frac{4s^2}{s-1} + \frac{4s^2}{s} \int_1^\infty u^{1-1/s} e^{-u}\, du$$
$$\leq (4s^2)\left(\frac{1}{s-1} + \frac{1}{s}\right) \leq 12s.$$

Thus, $\sup_x x^2|4L(2x) - L(x)| \leq 24s$, where we are making no attempt to optimize the bounds. Thus, the bound derived in Section 11.3 for Lipschitz kernels is applicable to ϕ with $C = \xi$ and $D = 24s$, so that

$$\kappa_n \leq 1 + 32 \times 24^2 \xi s^2 n^3. \quad \square$$

The optimization procedure goes as follows. Find Riemann approximations K_s' for all kernels K_s in our class $(\mathcal{K})$ with s even and $s \leq 2n^3$, making sure that $\int |K_s - K_s'| \leq 1/n$ for all these s. Note that uniformly over $s \leq 2n^3$, $\kappa_n \leq 1 + 73728\xi n^9 \leq \chi n^9$ with $\chi = 199804 > 1 + 73728\xi$. Let $\Theta = \{(h, s) : h > 0, 2 \leq s \leq 2n^3, s \text{ even}\}$, and pick the best pair (H, S) from Θ as in the previous section, where we picked the best Riemann kernel from a finite class of size N, together with an associated bandwidth. Then the estimate is $f_{n-m,K_S,H}$. The method is computationally intensive as the Riemann approximants require a lot of work. However, neither s nor h are a priori restricted, so this is a truly universal method. The performance is bounded uniformly over all densities by the following inequality:

THEOREM 16.3. *For the method described above, which picks (H, S) from Θ, we have for all n, $m \leq n/2$, and f:*

$$\mathbf{E}\left\{\int |f_{n-m,K_S,H} - f|\right\}$$
$$\leq 3 \inf_{h>0, s\geq 2} \mathbf{E}\left\{\int |f_{n,K_s,h} - f|\right\}\left(1 + \frac{2m}{n-m} + 8\sqrt{\frac{m}{n}}\right)$$
$$+ 8\sqrt{\frac{26 + 29\log n}{m}} + \frac{1053}{n}.$$

PROOF. By Theorem 2, if we merely look at the class of kernels K_s with $s \leq 2n^3$, we have

$$\mathbf{E}\left\{\int |f_{n-m,K_S,H} - f|\right\}$$
$$\leq 3 \inf_{h>0, 2\leq s\leq 2n^3} \mathbf{E}\left\{\int |f_{n,K_s,h} - f|\right\}\left(1 + \frac{2m}{n-m} + 8\sqrt{\frac{m}{n}}\right)$$
$$+ 8\sqrt{\frac{\log(9/2\, n^{29}\chi^2)}{m}} + \frac{30}{n},$$

where we replaced N by n^3, and κ_n by its uniform upper bound χn^9. To extend the infimum on the left-hand side of this inequality over all s, we note the following:

$$\inf_{h>0,2\le s\le 2n^3} \mathbf{E}\left\{\int |f_{n,K_s,h}-f|\right\}$$
$$\le \inf_{h>0,2\le s} \mathbf{E}\left\{\int |f_{n,K_s,h}-f|\right\}$$
$$+ \sup_{h>0,s>2n^3} \left(\mathbf{E}\left\{\int |f_{n,K_s,h}-f|\right\} - \mathbf{E}\left\{\int |f_{n,K_{2n^3},h}-f|\right\}\right)$$
$$\le \inf_{h>0,2\le s} \mathbf{E}\left\{\int |f_{n,K_s,h}-f|\right\} + \sup_{s>2n^3} \int |K_s - K_{2n^3}|$$
$$\le \inf_{h>0,2\le s} \mathbf{E}\left\{\int |f_{n,K_s,h}-f|\right\} + 2\sup_{s\ge 2n^3} \int |K_s - K_\infty|,$$

where K_∞ is the limit kernel corresponding to ϕ_∞. By Lemma 2, $|K_s(x)| \le 72s/x^2$, while from the expression for K_∞, $|K_\infty(x)| \le 4/(\pi x^2)$, so

$$\int_{|x|>r} |K_s - K_\infty| \le \frac{48s + 8/\pi}{r}.$$

Furthermore,

$$\int_{|x|\le r} |K_s - K_\infty| \le \sqrt{2r}\sqrt{\int |K_s - K_\infty|^2}$$

(by the Cauchy–Schwartz inequality)

$$= \sqrt{r/\pi}\sqrt{\int |\phi_s - \phi_\infty|^2}$$

(by Parseval's identity)

$$\le \sqrt{r/\pi}\sqrt{18\int |\psi_s - \psi_\infty|^2}$$

(by the c_r inequality, where $\psi_\infty(t) = (1-|t|)_+$).

For $0 \le t \le 1$, we have

$$|\psi_s(t) - \psi_\infty(t)| = (1-t)(1-e^{-t^s}) \le (1-t)t^s \le \left(\frac{s}{s+1}\right)^s \frac{1}{s+1} < \frac{1}{s+1}.$$

For $1 \le t \le 1 + 1/s$, we have

$$|\psi_s(t) - \psi_\infty(t)| = (t-1)e^{-t^s} < \frac{1}{s}.$$

For $t \ge 1 + 1/s$, the derivative of $(t-1)e^{-t^s}$ is nonpositive, so that $\sup_t |\psi_s - \psi_\infty| \le 1/s$. In particular, then,

$$\int |\psi_s - \psi_\infty|^2 \le \frac{4}{s^2} + 2\int_2^\infty (t-1)^2 e^{-2t^s}\, dt$$

$$\le \frac{4}{s^2} + 2e^{-2^s} \int_2^\infty t^2 e^{-t^2}\, dt$$
$$\le \frac{4}{s^2} + e^{-2^s} < \frac{5}{s^2}.$$

Combining all this shows that

$$\begin{aligned}\int |K_s - K_\infty| &\le \frac{48s + 8/\pi}{r} + \sqrt{\frac{90r}{\pi s^2}} < \frac{50s}{r} + \sqrt{\frac{30r}{s^2}} \\ &= \frac{10}{s^{1/3}} + \frac{\sqrt{150}}{s^{1/3}} \qquad \text{(take } r = 5s^{4/3}\text{)} \\ &< \frac{23}{s^{1/3}} \le \frac{19}{n}\end{aligned}$$

as $s \ge 2n^3$. Thus,

$$\inf_{h>0, 2\le s\le 2n^3} \mathbf{E}\left\{\int |f_{n,K_s,h} - f|\right\} \le \inf_{h>0,2\le s} \mathbf{E}\left\{\int |f_{n,K_s,h} - f|\right\} + \frac{38}{n}.$$

Finally,

$$\begin{aligned}&\mathbf{E}\left\{\int |f_{n-m,K_S,H} - f|\right\} \\ &\le 3 \inf_{h>0,2\le s} \mathbf{E}\left\{\int |f_{n,K_s,h} - f|\right\} \left(1 + \frac{2m}{n-m} + 8\sqrt{\frac{m}{n}}\right) \\ &\quad + 8\sqrt{\frac{\log\left(\tfrac{9}{2} n^{29} \chi^2\right)}{m}} + \frac{30}{n} + \frac{3 \times 38 \times 9}{n}.\end{aligned}$$

Theorem 3 now follows as $\log(9\chi^2/2) < 26$. □

The constants in Theorem 3 are of course not best possible, but we only wanted to work out one example in all its details, to show that the algorithm has indeed the required asymptotic performance. Undoubtedly, in practice, one either uses one infinite order kernel or optimizes over just a few low-order kernels. As shown above, nothing keeps one from optimizing over an infinite class of kernels. In fact, it is of independent interest to design such infinite classes with computational efficiency in mind.

§16.5. Bibliographic Remarks

The lower bound for the expected error with positive kernels is due to Devroye and Penrod (1984). The fact that negative-valued kernels can, in some cases, yield rates of convergence faster than $n^{-2/5}$ has been known for a long time (Bartlett, 1963). Berlinet (1993) surveys kernel classes and shows how to construct of hierarchies of kernels with various optimality properties. The value of $\Phi(n, K)$ in an L_2 setting were studied and obtained by Watson and Leadbetter (1963) (see also Davis (1975, 1977)). For L_1, only bounds are known (Devroye, 1988).

§16.6. Exercises

Exercise 16.1 (KERNELS IN $\mathcal{K}_4$). Examples of such kernels include the following:

A. The Granovsky–Müller kernel $(1/32)(45-150x^2+105x^4)\mathbf{1}_{[-1,1]}(x)$ (Granovsky and Müller, 1989).
B. The Müller kernel $(105/64)(1-5x^2+7x^4-3x^6)\mathbf{1}_{[-1,1]}(x)$ (Müller, 1984).
C. The Gasser–Müller–Mammitzsch kernel $(75/16)(1-x^2)_+ - (105/32)(1-x^4)_+$ (Gasser, Müller, and Mammitzsch, 1985).
D. The kernel $(1/8)(9-15x^2)\mathbf{1}_{[-1,1]}(x)$.

Asymptotic analysis of the expected L_1 error (Devroye, 1987) suggests that if these kernels K are rescaled such that $\int x^4 K(x)\,dx = 1$, then the best performance is obtained by the kernel that minimizes $\int K^2$. Which one of the above kernels is best in this sense?

Exercise 16.2 Stuetzle and Mittal (1979) suggested a method to construct a kernel in $\mathcal{K}_{2s}$ from a kernel in $\mathcal{K}_s$. Let s be an even positive integer, and let $K \in \mathcal{K}_s$. Then show that $L = 2K - K * K \in \mathcal{K}_{2s}$.

Exercise 16.3 Let f be a density having $s-1$ absolutely continuous derivatives, such that $f^{(s)}$ is continuous, and that $\int |f^{(s)}| < \infty$. Show that the characteristic function φ of f satisfies the inequality

$$\sup_t |t|^s |\varphi(t)| \le \int |f^{(s)}| < \infty.$$

Exercise 16.4 Find expressions for the densities that correspond to the Hall–Marron kernels that have characteristic functions $e^{-|t|^s}$, with s positive and even (the case $s = 2$ is easy).

§16.7. References

M. S. Bartlett, "Statistical estimation of density functions," *Sankhya Series A*, vol. 25, pp. 245–254, 1963.

A. Berlinet, "Hierarchies of higher order kernels," *Probability Theory and Related Fields*, vol. 94, pp. 489–504, 1993.

K. B. Davis, "Mean square error properties of density estimates," *Annals of Statistics*, vol. 5, pp. 1025–1030, 1975.

K. B. Davis, "Mean integrated square error properties of density estimates," *Annals of Statistics*, vol. 5, pp. 530–535, 1977.

L. Devroye, "Asymptotic performance bounds for the kernel estimate," *Annals of Statistics*, vol. 16, pp. 1162–1179, 1988.

L. Devroye, "A universal lower bound for the kernel estimate," *Statistics and Probability Letters*, vol. 8, pp. 419–423, 1989.

L. Devroye, "Universal smoothing factor selection in density estimation: theory and practice (with discussion)," *Test*, vol. 6, pp. 223–320, 1997.

L. Devroye, *A Course In Density Estimation*, Birkhäuser-Verlag, Boston, 1987.

L. Devroye and L. Györfi, *Nonparametric Density Estimation. The L_1 View*, Wiley, New York, 1985.

L. Devroye and C. S. Penrod, "Distribution-free lower bounds in density estimation," *Annals of Statistics*, vol. 12, pp. 1250–1262, 1984.

T. Gasser, H.-G. Müller, and V. Mammitzsch, "Kernels for nonparametric curve estimation," *Journal of the Royal Statistical Society, Series B*, vol. 47, pp. 238–252, 1985.

B. L. Granovsky and H.-G. Müller, "On the optimality of a class of polynomial kernel functions," *Statistics and Decisions*, vol. 7, pp. 301–312, 1989.

P. Hall and J. S. Marron, "Choice of kernel order in density estimation," *Annals of Statistics*, vol. 16, pp. 161–173, 1988.

H.-G. Müller, "Smooth optimum kernel estimators of densities, regression curves and modes," *Annals of Statistics*, vol. 12, pp. 766–774, 1984.

W. Stuetzle and Y. Mittal, "Some comments on the asymptotic behavior of robust smoothers," in: *Proceedings of the Heidelberg Workshop* (edited by T. Gasser and M. Rosenblatt), pp. 191–195, Springer Lecture Notes in Mathematics 757, Springer-Verlag, Heidelberg, 1979.

G. S. Watson and M. R. Leadbetter, "On the estimation of the probability density," *Annals of Mathematical Statistics*, vol. 34, pp. 480–491, 1963.

Chapter 17
Bandwidth Choice with Superkernels

§17.1. Superkernels

In the previous chapter, we looked at the saturation classes for kernels with many vanishing moments. The best rate $\Phi(n, K)$ for these kernels was $n^{-s/(2s+1)}$, with s an even positive integer. So, in theory, if we could find a kernel all of whose moments are vanishing, then we could hope for a rate $\Phi(n, K)$ that is better than $n^{-1/2+\epsilon}$ for all $\epsilon > 0$. So, is it possible to find such kernels? Can we simultaneously insure $\int K = 1$, yet $\int x^i K(x)\,dx = 0$ for all $i > 0$? The answer is yes. In fact, there are many such kernels, and each of them has of course its own best rate $\Phi(n, K)$. This could lead to the introduction of yet another hierarchy of saturation classes with best rates, for example, $(\log n)^s/\sqrt{n}$. However, as these saturation classes are surely becoming rather small—all densities in them must be infinitely many times continuously differentiable—this is an almost pointless exercise. Instead, we consider the ultimate kernels, those for which $\Phi(n, K) = O(1/\sqrt{n})$. Clearly, $O(1/\sqrt{n})$ is the best possible rate, as Lemma 1 shows.

LEMMA 17.1 (DEVROYE, 1989). *Let $f_{n,K,h}$ be the standard kernel estimate on* **R** *with kernel K and bandwidth h. Then*

$$\inf_{h.K,f} \mathbf{E}\left\{\int |f_{n,K,h} - f|\right\} \geq \frac{1}{\sqrt{528n}}.$$

PROOF. Let φ be the characteristic function for f, and let ψ be the Fourier transform of K_h defined by $\psi(t) = \int e^{itx} K_h(x)\,dx$. Observe that the convolution $f * K_h$ has Fourier transform $\varphi(t)\psi(t)$. The dependence upon h is absorbed in ψ. From simple inequalities (see the exercises and the proof of Theorem 9.5), we have, writing f_n for $f_{n,K,h}$,

$$\mathbf{E}\left\{\int |f_n - f|\right\} \geq \int |f - f * K_h| \geq \sup_t |\varphi(t) - \varphi(t)\psi(t)| = \sup |\varphi||1 - \psi|.$$

Also,

$$\mathbf{E}\left\{\int |f_n - f|\right\} \geq \tfrac{1}{2}\mathbf{E}\left\{\int |f_n - f * K_h|\right\}$$

$$\geq \tfrac{1}{2}\mathbf{E}\left\{\sup_t |\varphi_n - \varphi||\psi|\right\}$$
$$\geq \tfrac{1}{2}\sup_t \mathbf{E}\{|\varphi_n - \varphi||\psi|\},$$

where $\varphi_n = (1/n)\sum_{j=1}^n e^{itX_j}$ is the empirical characteristic function. Note that we used the fact that the Fourier transform of f_n is $\varphi_n\psi$. Next, $|\varphi_n - \varphi| \geq |\Re(\varphi_n) - \Re(\varphi)|$, where $\Re(z)$ denotes the real part of a complex number z. A similar inequality holds true for the imaginary part. But

$$\Re(\varphi_n(t)) - \Re(\varphi(t)) = \frac{1}{n}\sum_{j=1}^n (\cos(tX_j) - \Re(\varphi(t)))$$

is an average of n i.i.d. zero mean random variables, whose absolute value does not exceed 2. By Lemma 10.3:

$$\begin{aligned}\sqrt{n}\mathbf{E}\{|\Re(\varphi_n) - \Re(\varphi)|\} &\geq \frac{1}{\sqrt{8}}\mathbf{E}\{|\cos(tX_1) - \Re(\varphi(t))|\}\\ &\geq \frac{1}{\sqrt{32}}\mathbf{E}\{(\cos(tX_1) - \Re(\varphi(t)))^2\}\\ &= \frac{1}{\sqrt{32}}\left(\tfrac{1}{2} + \frac{\Re(\varphi(2t))}{2} - \Re^2(\varphi(t))\right).\end{aligned}$$

A similar argument on the imaginary part can be used to show that

$$\sqrt{n}\mathbf{E}\{|\Im(\varphi_n) - \Im(\varphi)|\} \geq \frac{1}{\sqrt{32}}\left(\tfrac{1}{2} - \frac{\Re(\varphi(2t))}{2} - \Im^2(\varphi(t))\right).$$

Averaging the two bounds yields

$$\sqrt{n}\mathbf{E}\left\{|\varphi_n - \varphi|\right\} \geq \frac{1}{\sqrt{128}}(1 - |\varphi|^2).$$

Collecting all the bounds shows that

$$\begin{aligned}\mathbf{E}\left\{\int |f_n - f|\right\} &\geq \max\left(\sup_t |1-\psi||\varphi|,\ \sup_t |\psi| n^{-1/2}\frac{1}{\sqrt{128}}(1 - |\varphi|^2)\right)\\ &\geq \sup_t \max\left(|1-\psi||\varphi|,\ |\psi| n^{-1/2}\frac{1}{\sqrt{128}}(1 - |\varphi|^2)\right)\\ &\geq \tfrac{1}{2}\sup_t \min\left(|\varphi|,\ n^{-1/2}\frac{1}{\sqrt{128}}(1 - |\varphi|^2)\right).\end{aligned}$$

The minimum is maximal when both operands are equal, which occurs for

$$|\varphi| = \tfrac{1}{2}(-\sqrt{128n} + \sqrt{128n + 4}) = \tfrac{1}{2}\sqrt{128n}\left(-1 + \sqrt{1 + \frac{4}{128n}}\right)$$

$$\geq \tfrac{1}{2}\sqrt{128n}\left(-1 + 1 + \frac{4}{256n\sqrt{1 + 4/128n}}\right)$$

$$\left(\text{use } \sqrt{1+\xi} \geq 1 + \frac{\xi}{2\sqrt{1+\xi}}\right)$$

$$= \frac{1}{\sqrt{128n}\sqrt{1+1/32n}}.$$

The value for $|\varphi|$ is between zero and one, so that, by the continuity of φ, it is attained for some t. The sought lower bound is ½ times the given value of $|\varphi|$. □

Kernels with $\Phi(n, K) = O(1/\sqrt{n})$ will be called superkernels. These kernels do not limit the expected L_1 error of the kernel estimate in any way, and in fact, for many densities f, if K is a superkernel,

$$\inf_h \mathbf{E}\left\{\int |f_{n,K,h} - f|\right\} \leq (c + o(1)) \inf_{h,L} \mathbf{E}\left\{\int |f_{n,L,h} - f|\right\},$$

where $c > 0$ is a constant. In other words, there is no need to optimize with respect to the pair (h, L) any more, as it suffices to just take K and optimize with respect to h, which is a much easier problem, computationally and complexity-wise.

In this chapter, we work out one such example. First, we identify a superkernel K, and show how to create others. Finally, in order to apply the minimum distance method of Chapter 11, it is necessary to find the kernel complexity of this superkernel. The technical hurdle here is caused by the fact that such superkernels have infinite support and oscillate infinitely often. We will be able to show that with the minimum distance choice H:

$$\mathbf{E}\left\{\int |f_{n,K,H} - f|\right\} \leq (3 + o(1)) \inf_h \mathbf{E}\left\{\int |f_{n,K,h} - f|\right\} + O\left(\sqrt{\frac{\log n}{n}}\right)$$

provided we take $m \sim n/\log n$.

§17.2. The Trapezoidal Kernel

Early work by Watson and Leadbetter (1963) and Davis (1975) suggested by studying the expected L_2 error that one should use the sinc kernel $S(x) = \sin x/\pi x$. The characteristic function of S is $\psi(t) = \mathbf{1}_{[-1,1]}(t)$. The flatness of ψ near the origin implies that the best L_2 error, $\inf_h \mathbf{E}\{\int (f_{nh} - f)^2\}$, adapts itself to the density at hand and is thus not a priori limited by the saturation phenomenon seen for finite-order kernels. Unfortunately, the sinc kernel is not absolutely integrable. To obtain an absolutely integrable yet comparable kernel, consider first the trapezoidal characteristic function (Devroye and Györfi, 1985):

$$\psi(t) = 2(1 - |t|/2)_+ - (1 - |t|)_+,$$

a difference between two triangular characteristic functions. As $(1-|t|)_+$ is the characteristic function for the de la Vallée-Poussin density

$$v(x) = \frac{1}{2\pi}\left(\frac{\sin(x/2)}{x/2}\right)^2,$$

we see that the kernel for ψ is

$$K(x) = 4v(2x) - v(x) = \frac{7+\cos x - 8\cos^2 x}{4\pi x^2}.$$

We call K the trapezoidal kernel. Note that since

$$\sup_x |K(x)| \le (1/2\pi)\int |\psi(t)|\,dt \le 2/\pi,$$

K is bounded and absolutely integrable, as required:

$$|K(x)| \le \frac{2}{\pi}\min\left(\frac{1}{x^2}, 1\right).$$

Furthermore, $\int K(x)\,dx = \psi(0) = 1$. Since $\psi(t) = \int e^{itx}K(x)\,dx$, we have for integer $s > 0$,

$$\psi^{(s)}(t) = \int (ix)^s e^{itx} K(x)\,dx,$$

and, in particular, $\psi^{(s)}(0) = \int (ix)^s K(x)\,dx = 0$. This shows that for any superkernel, it is necessary to have a characteristic function with zero derivatives at the origin. The trapezoidal kernel K thus fits the bill.

§17.3. Bandwidth Selection

The trapezoidal kernel is a candidate for our minimum distance method because it is polynomially Riemann approximable, even though it oscillates infinitely often:

LEMMA 17.2. *The trapezoidal kernel is polynomially Riemann approximable and has kernel complexity κ_n not exceeding $51n^3$.*

PROOF. We may apply the bound of Section 11.3 for the kernel complexity of Lipschitz kernels here. Indeed, standard calculations and the obvious inequalities $|x - \sin x| \le |x|^3/6$ and $|\sin x| \le |x|$ allow us to verify that v is Lipschitz and that in fact $\sup_x |v'(x)| \le 1/(3\pi)$. Thus, $\sup_x |K'(x)| \le \sup_x 8|v'(2x)| + \sup_x |v'(x)| \le 9/(3\pi) = 3/\pi$. Furthermore, $|K(x)| \le 4/(\pi x^2)$. Then, we note that the cited bound of Section 11.3 gives

$$\kappa_n \le 1 + 32(3/\pi)(4/\pi)^2 n^3 = 1 + \frac{1536n^3}{\pi^3} < 51n^3. \quad \square$$

Lemma 2 provides us, furthermore, with a useful construction for carrying out the computations, as the approximating finite-order Riemann kernel is explicitly constructed. Thus, taking $m = n/2$, for example, we see that for the minimum distance estimate $f_{n-m,K,H}$ defined in Chapter 11:

$$\mathbf{E}\left\{\int |f_{n-m,K,H} - f|\right\} \leq 26 \inf_{h>0} \mathbf{E}\left\{\int |f_{n,K,h} - f|\right\} + D\sqrt{\frac{\log n}{n}}$$

for a universal constant D. For the majority of densities, the last term in the bound is asymptotically negligible with respect to the first term. And, as is well known, the trapezoidal kernel is optimal to within a multiplicative factor within the class of all kernels (see Devroye (1988, 1992) for an L_1 treatment of this subject). Thus, we have a method here for picking h and K (the trapezoidal kernel) such that

$$\mathbf{E}\left\{\int |f_{n,K,H} - f|\right\} \leq C \inf_{L,h} \mathbf{E}\left\{\int |f_{n,L,h} - f|\right\} + D\sqrt{\frac{\log n}{n}}$$

for constants C and D. In other words, for most densities, we have reached the limit of what is possible with standard kernel estimates. Densities for which the second term is asymptotically larger include the normal density, where the first term is $O((\log n)^{1/4}/\sqrt{n})$, and all densities for which the characteristic function has compact support.

§17.4. Bibliographic Remarks

Watson and Leadbetter (1963) and later, Konakov (1972), Davis (1975, 1977), Ibragimov and Khasminskii (1982), Devroye (1987, 1988, 1992), and many others documented the usefulness of infinite order kernels K in the Parzen–Rosenblatt estimate when the dimension, d, is one.

§17.5. Exercises

Exercise 17.1 Show that if f is an integrable function with Fourier transform $\psi(t) = \int e^{itx} f(x)\, dx$, then

$$\int |f| \geq \sup_t |\psi(t)|$$

and

$$\int |\psi(t)|\, dt \geq 2\pi \sup_x |f(x)|.$$

Exercise 17.2 (NATTERER'S KERNEL (PAWLAK AND STADTMÜLLER, 1996)). Let K be the kernel with characteristic function

$$\psi(t) = e^{-1/(1-t^2)} \mathbf{1}_{[-1,1]}(t).$$

Show that $|K(x)| \leq c_1 e^{-c_2\sqrt{x}}$ for $x > 1$ and some positive constants c_1, c_2. Estimate the kernel complexity for ψ. Show that all moments of K are zero and that $\int K = 1$.

Exercise 17.3 Let $\psi(t)$ be a characteristic function of a kernel K, satisfying $\psi(t) = 1$ for $|t| \leq c$, where $c > 0$ is fixed. Show that there is an infinite sequence $0 < x_1 < x_2 < x_3 < \cdots$ such that $K(x_i)$ is negative for odd i and positive for even i.

Exercise 17.4 Consider the superkernel K with characteristic function

$$\psi(t) = \begin{cases} 1 & |t| \leq 1 \,, \\ \frac{1}{1+(|t|-1)^4} & |t| > 1 \,. \end{cases}$$

A. Show that $|K|$ and $x^2|K(x)|$ are bounded.
B. Show that $\int_z^\infty |K(x)|\,dx \leq c/z$ for some constant c and all $z > 0$.
C. Show that the kernel complexity of K is polynomial in n, and find an explicit upper bound for it.

Exercise 17.5 Consider the normal density with characteristic function $\varphi(t) = \exp(-t^2)$. Estimate it by using the kernel estimate f_{nh} with the kernel of the previous exercise. Let the bandwidth h be a function of n such that $h \to 0$, $nh \to \infty$.

A. Show that $\mathbf{E}\{\int |f_{nh} - f * K_h|\} \leq c_1/\sqrt{nh}$ for some positive constant c_1.
B. Show that $\sup_x |f(x) - f * K_h(x)| \leq e^{-(1+o(1))/h^2}$.
C. Using the above facts, and tail bounds for K (from the previous exercise) and f, show that $\int |f - f * K_h| \leq c_2 e^{-c_3/h^2}$ for some positive constants c_i.
D. Show that one can find a sequence h such that

$$\int |f_{nh} - f| \leq \frac{C \log^{1/4} n}{\sqrt{n}}$$

for some constant C. Incidentally, this is the best possible rate for the normal density achievable with any kernel (Devroye, 1988).

Exercise 17.6 In this exercise, we identify the saturation class for superkernels with a flattop characteristic function. Consider the trapezoidal kernel K. Let f be a density whose characteristic function vanishes off $[-T, T]$. Show that there exists a threshold h^*, such that for all $h < h^*$, $f \equiv f * K_h$, so that the bias in the L_1 error is zero. Then show that if f has a finite second moment,

$$\sup_{h<h^*} \sqrt{nh}\,\mathbf{E}\left\{\int |f_{nh} - f * K_h|\right\} \leq c$$

for some positive constant c, where f_{nh} is the kernel estimate with kernel K and bandwidth h. Conclude that if we keep h fixed and less than h^*, then $\mathbf{E}\{\int |f_{nh} - f * K_h|\} = O(1/\sqrt{n})$.

Exercise 17.7 Find a density f with characteristic function vanishing off $[-T, T]$ and for which all moments are finite. If you cannot do that, find one for which the fourth moment is finite.

Exercise 17.8 Cline (1990) has shown that if $f_{n,K,h}$ is the standard kernel estimate on $\mathbf{R}$ with Epanechnikov kernel K and bandwidth h, then there exists another kernel L such that for all n and h, $\mathbf{E}\{\int (f_{n,L,h} - f)^2\} \leq \mathbf{E}\left\{\int (f_{n,K,h} - f)^2\right\}$, the inequality being strict for most densities. It suffices to give L the characteristic function $\max(\psi, 0)$, where ψ is the characteristic function for K. In other words, the Epanechnikov kernel is inadmissible. While the operation $\max(\psi, 0)$ improves the expected L_2 error of any symmetric kernel K, not just the Epanechnikov kernel, explore whether the same thing is true for the L_1 error, by either establishing the L_1 version of the inequality or providing a counterexample.

§17.6. References

D. B. H. Cline, "Optimal kernel estimation of densities," *Annals of the Institute of Statistical Mathematics*, vol. 42, pp. 287–303, 1990.

K. B. Davis, "Mean square error properties of density estimates," *Annals of Statistics*, vol. 5, pp. 1025–1030, 1975.

K. B. Davis, "Mean integrated square error properties of density estimates," *Annals of Statistics*, vol. 5, pp. 530–535, 1977.

L. Devroye, *A Course in Density Estimation*, Birkhauser-Verlag, Boston, 1987.

L. Devroye, "Asymptotic performance bounds for the kernel estimate," *Annals of Statistics*, vol. 16, pp. 1162–1179, 1988.

L. Devroye, "A universal lower bound for the kernel estimate," *Statistics and Probability Letters*, vol. 8, pp. 419–423, 1989.

L. Devroye, "A note on the usefulness of superkernels in density estimation," *Annals of Statistics*, vol. 20, pp. 2037–2056, 1992.

L. Devroye and L. Györfi, *Nonparametric Density Estimation: The L_1 View*, Wiley, New York, 1985.

I. A. Ibragimov and R. Z. Khasminskii, "Estimation of distribution density belonging to a class of entire functions," *Theory of Probability and its Applications*, vol. 27, pp. 551–562, 1982.

V. D. Konakov, "Non-parametric estimation of density functions," *Theory of Probability and its Applications*, vol. 17, pp. 361–362, 1972.

M. Pawlak and U. Stadtmüller, "Recovering band-limited signals under noise," *IEEE Transactions on Information Theory*, vol. IT-42, pp. 1425–1438, 1996.

G. S. Watson and M. R. Leadbetter, "On the estimation of the probability density, I," *Annals of Mathematical Statistics*, vol. 34, pp. 480–491, 1963.

Author Index

Subject Index

Springer Series in Statistics

(continued from p.ii)

Mosteller/Wallace: Applied Bayesian and Classical Inference: The Case of the Federalist Papers.
Parzen/Tanabe/Kitagawa: Selected Papers of Hirotugu Akaike.
Politis/Romano/Wolf: Subsampling.
Ramsay/Silverman: Functional Data Analysis.
Rao/Toutenburg: Linear Models: Least Squares and Alternatives.
Read/Cressie: Goodness-of-Fit Statistics for Discrete Multivariate Data.
Reinsel: Elements of Multivariate Time Series Analysis, 2nd edition.
Reiss: A Course on Point Processes.
Reiss: Approximate Distributions of Order Statistics: With Applications to Non-parametric Statistics.
Rieder: Robust Asymptotic Statistics.
Rosenbaum: Observational Studies.
Rosenblatt: Gaussian and Non-Gaussian Linear Time Series and Random Fields.
Särndal/Swensson/Wretman: Model Assisted Survey Sampling.
Schervish: Theory of Statistics.
Shao/Tu: The Jackknife and Bootstrap.
Siegmund: Sequential Analysis: Tests and Confidence Intervals.
Simonoff: Smoothing Methods in Statistics.
Singpurwalla and Wilson: Statistical Methods in Software Engineering: Reliability and Risk.
Small: The Statistical Theory of Shape.
Sprott: Statistical Inference in Science.
Stein: Interpolation of Spatial Data: Some Theory for Kriging.
Taniguchi/Kakizawa: Asymptotic Theory of Statistical Inference for Time Series.
Tanner: Tools for Statistical Inference: Methods for the Exploration of Posterior Distributions and Likelihood Functions, 3rd edition.
Tong: The Multivariate Normal Distribution.
van der Vaart/Wellner: Weak Convergence and Empirical Processes: With Applications to Statistics.
Verbeke/Molenberghs: Linear Mixed Models for Longitudinal Data.
Weerahandi: Exact Statistical Methods for Data Analysis.
West/Harrison: Bayesian Forecasting and Dynamic Models, 2nd edition.